# Pearson
# Revise

KU-631-507

## AQA GCSE (9–1)
# Mathematics
## Foundation tier
## Revision Guide

Series Consultant: Harry Smith

Author: Harry Smith

## Also available to support your revision:

Revise GCSE Study Skills Guide      9781292318875

The **Revise GCSE Study Skills Guide** is full of tried-and-trusted hints and tips for how to learn more effectively. It gives you techniques to help you achieve your best – throughout your GCSE studies and beyond!

Revise GCSE Revision Planner      9781292318868

The **Revise GCSE Revision Planner** helps you to plan and organise your time, step-by-step, throughout your GCSE revision. Use this book and wall chart to mastermind your revision.

**Question difficulty**
Look at this scale next to each exam-style question. It tells you how difficult the question is.

Worked solution video

**Worked solution videos**
Some of the questions in this Revision Guide have worked solution videos. Use your phone, tablet or webcam to scan the QR code to watch the video.

**For the full range of Pearson revision titles across KS2, 11+, KS3, GCSE, Functional Skills, AS/A Level and BTEC visit:**
www.pearsonschools.co.uk/revise

# Contents

**A small bit of small print**
AQA publishes Sample Assessment Material and the Specification on its website. This is the official content and this book should be used in conjunction with it. The questions in *Now try this* have been written to help you practise every topic in the book. Remember: the real exam questions may not look like this.

# Place value

The value of each digit in a number depends on its position. Digits that are further to the left are worth more. You can use a place value diagram to help you read and write numbers.

## Worked example

Write these amounts in order, smallest first.

**(1 mark)**

£2908    £2950    £5011    £925    £10 430

£925
£925                          £10 430
£925                  £5011   £10 430
£925   £2908   £2950   £5011   £10 430

**Check it!**
Have you included all the amounts in your final answer? ✓

## Working with money

- ✓ Do all your calculations in the same units, either £ or p.
- ✓ Write either £ or p in your answer, but not both.
- ✓ 100 p = £1
- ✓ Amounts in pounds need 2 decimal places. Write 280 p as £2.80.

## Worked example

A music website sells songs and albums.
Songs cost 79p each.
Albums cost £6.99 each.
Aaron has a £25 gift card.
He buys 2 albums and spends the rest on songs.
How many songs can he afford?    **(2 marks)**

This scale tells you how hard the question is.

2 × 6.99 = 13.98
25 − 13.98 = 11.02
11.02 ÷ 0.79 = 13.949...
Aaron can buy 13 songs.

## Problem solved!

There is more than one successful strategy for a question like this. Whichever strategy you choose, you have to show your working.
Choose whether to work in pounds or pence.
If you divide the amount of money Aaron has left by 79p you need to round **down** because Aaron can only buy a whole number of songs.

**Check it!**
2 × £6.99 + 13 × £0.79 = £24.25 ✓
2 × £6.99 + 14 × £0.79 = £25.04 ✗

You will need to use problem-solving skills throughout your exam – **be prepared!**

## Now try this

**1** Write these amounts in order, smallest first.
£1974   £974   £1749   £1497   £1947

**(1 mark)**

**Profit** is money you make, and **loss** is money you lose. You need to **add** the individual profits, then **subtract** the individual losses.

**2** Mark buys and sells used cars. The table shows information about five cars that Mark bought and sold.

| Car | A | B | C | D | E |
|---|---|---|---|---|---|
| Bought | £2200 | £7800 | £4200 | £........ | £6500 |
| Sold | £2900 | £9000 | £...... | £11200 | £5750 |
| Profit or loss | £700 profit | £...... ........ | £1500 Loss | £500 profit | £...... ....... |

(a) Complete the table.    **(2 marks)**
(b) Work out Mark's total profit or loss for these five cars.    **(2 marks)**

# Negative numbers

Numbers smaller than 0 are called **negative** numbers.

NEGATIVE NUMBERS | POSITIVE NUMBERS

$-5 \ -4 \ -3 \ -2 \ -1 \ 0 \ 1 \ 2 \ 3 \ 4 \ 5$

0 is neither positive nor negative.

You can use a number line to write numbers in order of size. The numbers get bigger as you move to the right.

## Number lines

You can use number lines to help when adding and subtracting.

$-6 + 4 = -2$          $4 - 5 = -1$

$-6 \ -5 \ -4 \ -3 \ -2 \ -1 \ 0 \ 1 \ 2 \ 3 \ 4 \ 5$

Always mark 0 on your number line.

## Adding and subtracting

To add or subtract a negative number, change the double signs first.

$+ \ - \ \rightarrow \ -$

$12 + -3 = 12 - 3 = 9$

$- \ - \ \rightarrow \ +$

$5 - -9 = 5 + 9 = 14$

### Golden rule

When you **add** a negative number the answer is **lower**.

When you **subtract** a negative number the answer is **higher**.

## Multiplying and dividing

When multiplying and dividing, use these rules to decide whether the answer will be positive or negative.

**1** If numbers have the **same** sign then the answer is **positive**.
$-3 \times -7 = 21$

**2** If numbers have **different** signs then the answer is **negative**.
$+80 \div -10 = -8$

### Problem solved!

One strategy is to try a few pairs and add up the totals. You could also work out what each total should be by adding together all 6 cards then dividing by 3:

$-4 + -3 + 0 + 2 + -6 + -1 = -12$

$-12 \div 3 = -4$

**Check it!**

Make sure you have used each card exactly once. ✓

You will need to use problem-solving skills throughout your exam – **be prepared!**

### Worked example

Here are six number cards.

$\boxed{-4} \ \boxed{-3} \ \boxed{0} \ \boxed{2} \ \boxed{-6} \ \boxed{-1}$

Sort the cards into three pairs so that each pair has the same total.    **(2 marks)**

$\boxed{-4}$ and $\boxed{0}$ Total = $-4$

$\boxed{-3}$ and $\boxed{-1}$ Total = $-4$

$\boxed{-6}$ and $\boxed{2}$ Total = $-4$

### Now try this

**1** (a) Work out $-2 + -4$
   (b) Work out $-1 - -6$
   **(2 marks)**

Worked solution video

**2** (a) Work out $-4 \times -12$
   (b) Work out $30 \div -5$
   **(2 marks)**

**3** Here are six number cards.

$\boxed{-10} \ \boxed{-3} \ \boxed{1} \ \boxed{2} \ \boxed{-9} \ \boxed{-5}$

Sort the cards into three pairs so that each pair has the same total.    **(2 marks)**

Try a few pairs and add up the totals.

# Rounding numbers

To **round** a number, you look at the next digit to the right on a place value diagram.

5 or more → round up, less than 5 → round down.

To round to the nearest 100, you look at the digit in the tens column. It is a 5, so round up. 3250 rounded to the nearest 100 is 3300.

To round to 1 decimal place (1 d.p.), you look at the digit in the second decimal place. It is a 4, so round down. 5.043 rounded to 1 d.p. is 5.0 You **need** to write the 0 to show that you have rounded to 1 d.p.

To round to the nearest whole number look at the digit in the tenths column. It is a 0, so round down. 5.043 rounded to the nearest whole number is 5.

## Significant figures

You always start counting **significant figures** from the first non-zero on the left.

27.05 rounded to 1 s.f. is 30 — The first non-zero digit is 2. The next digit is 7, so round up to give an answer of 30.

27.05 rounded to 2 s.f. is 27 — Look for the two digits furthest to the left, which are 2 and 7. The next digit is 0, so round down to give an answer of 27.

27.05 rounded to 3 s.f. is 27.1 — The first three significant figures are 2, 7 and 0. The next digit is 5 so round up to 27.1.

### Numbers less than 1

When rounding numbers less than 1 to a given number of significant figures, remember **not** to count zero digits that are on the left.

0.0085 rounded to 1 s.f. is 0.009 — Look for the digit which is furthest to the left and which is **NOT** a zero. This digit is 8. The next digit is 5 so round up to give an answer of 0.009

## Worked example

Round 6.9083 correct to
(a) 1 significant figure
(b) 2 significant figures
(c) 3 significant figures    **(3 marks)**

(a) 7 (1 s.f.)
(b) 6.9 (2 s.f.)
(c) 6.91 (3 s.f.)

The width of the cooker is given to the nearest whole unit. So it might be **inaccurate** by up to half a unit in either direction. Show Supraj is wrong by giving an example of a width larger than 76 cm that would round to 76 cm.

There is more about rounding errors on page 33.

## Now try this

 1 Round the number 3756
(a) to the nearest 100    **(1 mark)**
(b) to the nearest 10    **(1 mark)**

 2 Write these numbers correct to 1 significant figure.
(a) 48    **(1 mark)**
(b) 3025    **(1 mark)**
(c) 0.00939    **(1 mark)**
(d) 6.5    **(1 mark)**

Worked solution video

 3 Round 0.179554 to 3 significant figures.    **(1 mark)**

 4 Supraj is installing a kitchen. He orders a cooker that is 76 cm wide, to the nearest cm. He says: "If I leave a gap exactly 76 cm wide the cooker will definitely fit perfectly." Is Supraj correct? Justify your answer. **(1 mark)**

# Adding and subtracting

You need to be able to add and subtract numbers without a calculator.

## Mental methods

Try these methods for adding and subtracting quickly in your head.

213 + 79

+70    +9

213          283 292

Add the tens first
then the units.
213 + 79 = 292

152 − 63

+7  +30    +52

63 70    100      152

Count on in steps from 63 up to 152.
Add up the steps to work out the
difference between 152 and 63.
7 + 30 + 52 = 89
152 − 63 = 89

---

**Worked example**

Work out 285 + 56 + 1091

**(2 marks)**

```
  285
   56
+1091
─────
 1432
  2 1
```

1. **Always** add the units column first:
   5 + 6 + 1 = 12. Write down the 2 and carry the
   1 over to the tens column.
2. Add the tens column: 8 + 5 + 9 + 1 = 23.
   Make sure you include any numbers you carried
   over. Write down the 3 and carry the **2** over to
   the hundreds column.
3. Add the hundreds column: 2 + 0 + 2 = 4.
   Write down 4.
4. There is only one digit in the thousands column.
   Write this in your answer.

---

**Worked example**

Work out 418 − 62

**(2 marks)**

```
 ³4¹18
−  62
─────
  356
```

1. **Always** subtract the units column first:
   8 − 2 = 6.
   Remember it is (top number) − (bottom number).
2. Look at the tens column. 1 is smaller than 6 so you
   have to exchange 1 hundred for 10 tens. Change
   4 hundreds into 3 hundreds and 10 tens.
3. Now you can subtract the tens column: 11 − 6 = 5.
4. There is nothing to subtract in the hundreds
   column so write 3 in your answer.

---

**Now try this**

1 Work out
   (a) 503 + 1126 + 85    **(2 marks)**
   (b) 745 + 283    **(2 marks)**

2 Work out
   (a) 627 − 251    **(2 marks)**
   (b) 831 − 659    **(2 marks)**

3 Here are some numbers.
   221  492  605  552  274  334
   Write the numbers in pairs so that the
   **sum** of the numbers in each pair is the
   same.    **(2 marks)**

4 Joe buys a magazine costing £4.45 and
   two birthday cards costing £1.99 each.
   He pays with a £10 note.
   How much change will he receive?

**(3 marks)**

You can work in pence so you don't have to use
decimal numbers. Work out 445 + 199 + 199,
then subtract the result from 1000.
Remember to give units with your answer.

# Multiplying and dividing

You need to be able to multiply and divide numbers without a calculator.
For a reminder about multiplying and dividing by 10, 100 and 1000 have a look at page 62.

## Mental methods

Try these methods for multiplying and dividing quickly in your head.

$37 \times 8$

$30 \times 8 = 240$

$\underline{7 \times 8 = \phantom{0}56}$

$37 \times 8 = 296$

Split 37 into 30 and 7 then multiply each by 8. Add the answers to get the total.

$54 \div 6$

$6 \times \square = 54$

The answer is 9.

Try to find a multiplication fact using 6 with 54 as the answer.

Multiplying and dividing are **much easier** if you know your times tables up to $10 \times 10$.

## Worked example

Work out
(a) $49 \times 3$ **(2 marks)**

$$\begin{array}{r} 49 \\ \times \phantom{0}3 \\ \hline 147 \\ {\scriptstyle 2} \end{array}$$

(b) $36 \times 24$ **(3 marks)**

$$\begin{array}{r} 36 \\ \times \phantom{0}24 \\ \hline 144 \\ {\scriptstyle 2} \\ 720 \\ \hline 864 \end{array}$$

Remember that this is a **non-calculator** question. You need to show all your working.
1. Work out $36 \times 4$. (Answer = 144)
2. Work out $36 \times 20$. Write down **0** and then work out $36 \times 2$. (Answer = 72**0**)
3. Add the answers. (144 + 720 = 864)

Always multiply from right to left.
1. $9 \times 3 = 27$. Write down 7 and carry over **2** (2 tens).
2. $4 \times 3 = 12$. Add on the carry-over. $12 + 2 = 14$. Write down 14.

## Worked example

Work out $288 \div 9$    **(2 marks)**

$$\begin{array}{r} 32\phantom{0} \\ 9)\overline{288} \\ -27\phantom{0} \\ \hline 18 \\ -18 \\ \hline 0 \end{array}$$

You can use long division.
1. Does 9 divide into 2? No.
2. Does 9 divide into 28? Yes. $9 \times 3 = 27$ so 9 divides into 28 three times with remainder 1.
3. Does 9 divide into 18? Yes. $9 \times 2 = 18$ so 9 divides into 18 two times with no remainder.
Using short division the calculation would look like this:

$$\begin{array}{r} 3\,2 \\ 9)\overline{28^{1}8} \end{array}$$

## Now try this

**1** (a) Work out $72 \times 100$    **(1 mark)**
(b) Work out $256 \times 9$    **(1 mark)**
(c) Work out $29 \times 78$    **(2 marks)**

**2** (a) Work out $468 \div 3$    **(2 marks)**
(b) Work out $1032 \div 8$    **(2 marks)**

**3** There are 18 chocolate coins in a bag.
Paula buys 6 of these bags.
Paula has 7 grandchildren. She wants to give each of her grandchildren 15 coins.
Has she bought enough coins?    **(3 marks)**

Worked solution video

You need to answer 'yes' or 'no' to the question, **and** show your working to **justify** your answer.

# Decimals and place value

You can use a place value diagram to help you understand and compare decimal numbers.
Remember that decimal numbers with more digits are not necessarily bigger. Try writing extra Os so that all the numbers have the same number of decimal places.

The value of the 5 in this number is 5 hundredths.

0.76 is the same as 0.760
0.76 is bigger than 0.758
6 hundredths is bigger than 5 hundredths.

0.79 is smaller than 0.8 because the digit in the tenths place is smaller.

## Worked example

Write these numbers in order, smallest first:
0.43    0.425    0.4    0.48    0.459
**(2 marks)**

~~0.43~~  ~~0.425~~  ~~0.4~~  ~~0.48~~  ~~0.459~~
0.4    0.425
0.4    0.425    0.43
0.4    0.425    0.43    0.459
0.4    0.425    0.43    0.459    0.48

All these numbers have the same tenths digit. You need to look at the hundredths digit first.
0.4 is the same as 0.40 so this is the smallest number.
0.425 has the next smallest hundredths digit so this is the next number.

**Check it!**
Cross out each number to make sure you include them all in your final answer. ✓

## Worked example

Using the information that
58 × 71 = 4118
write down the value of
(a) 58 × 0.71        **(1 mark)**
41.18
(b) 5800 × 7.1        **(1 mark)**
41180

(a) 71 has been divided by 100 and 58 hasn't been changed. So the answer needs to be divided by 100.        4118 ÷ 100 = 41.18
(b) 58 has been multiplied by 100 and 71 has been divided by 10.

─┤×100├─┤÷10├→ is the same as ─┤×10├→

The answer needs to be multiplied by 10:
4118 × 10 = 41180
To revise multiplying and dividing by 10, 100 and 1000 see page 62.

## Now try this

1  Write down the place value of the 8 in these numbers.
   (a) 2.84   **(1 mark)**        (b) 0.3086   **(1 mark)**
2  Write these numbers in order of size, smallest first.
   0.517  0.508  0.58  0.571  0.51        **(2 marks)**

3  Using the information that
   672 × 13 = 8736
   write down the value of
   (a) 0.672 × 13        **(1 mark)**
   (b) 8736 ÷ 67.2        **(1 mark)**

# Operations on decimals

## 1 Adding and subtracting

To add or subtract decimal numbers:
1. Line up digits with the same place value.
2. Line up the decimal points.
3. Write a decimal point in your answer.

See page 4 for a reminder about adding and subtracting.

Write in 0s so that both numbers have the same number of decimal places.

### Worked example

(a) 0.75 + 1.6          (b) 3.5 − 0.21
**(1 mark)**                    **(1 mark)**

```
   0.75              3.⁴5̷⁰1
 + 1.60            − 0.21
   2.35              3.29
   ₁
```

## 2 Multiplying

To multiply decimal numbers:
1. Ignore the decimal points and just multiply the numbers.
2. Count the number of decimal places in the calculation.
3. Put this number of decimal places in the answer.

(a) You can use estimation to check that the decimal point is in the correct place.
$8.69 \times 12 \approx 9 \times 12 = 108$     $108 \approx 104$ ✓

(b) $8.5 \times 0.04$ has 3 decimal places in total

So $8.5 \times 0.04 = 0.3\,4\,0 = 0.34$

Write a 0 before the decimal point and simplify your answer.

### Worked example

(a) 8.69 × 12          (b) 8.5 × 0.04
**(2 marks)**                    **(2 marks)**

```
    869              85
  ×  12            ×  4
   1738             340
 + 8690              2
  10428      8.5 × 0.04 = 0.34
   ₁ ₁
8.69 × 12 = 104.28
```

## 3 Dividing

To divide by a decimal number:
1. Multiply both numbers by 10, 100 or 1000 to make the second number a whole number.
2. Divide by the whole number.

Multiply 40.6 and 1.4 by 10.
If you multiply both numbers in a division by the same amount, the answer stays the same.

### Worked example

(a) 55.8 ÷ 3          (b) 40.6 ÷ 1.4     **(2 marks)**
**(1 mark)**    40.6 ÷ 1.4 = 406 ÷ 14

```
   18.6                29
3)5²5.¹8           14)406
                    −28
                    126
                   −126
                      0
             40.6 ÷ 1.4 = 29
```

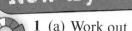

## Now try this

1  (a) Work out 12.5 + 7.93          **(2 marks)**
   (b) Work out 8.14 + 3 + 0.772     **(2 marks)**

2  (a) Work out 16.5 − 9.72          **(2 marks)**
   (b) Work out 5.76 × 34            **(2 marks)**

3  (a) A kitchen stool costs £39.90.
       Work out the cost of 6 kitchen stools.   **(2 marks)**
   (b) Tom buys 12 pencils for £11.28.
       Work out the cost of 1 pencil.           **(2 marks)**

Worked solution video

Make sure you line up the decimal points.

# Squares, cubes and roots

## Squares and square roots

When a number is multiplied by itself the answer is a **square number**. You can write square numbers using index notation.

| Multiplication | Index notation | Square number |
|---|---|---|
| 2 × 2 | $2^2$ | 4 |
| 5 × 5 | $5^2$ | 25 |
| 9 × 9 | $9^2$ | 81 |
| 13 × 13 | $13^2$ | 169 |

Square numbers are the areas of squares with whole number side lengths.
There is more about area on page 81.

25 cm² 5 cm
5 cm

**Square roots** are the opposite of squares. You use the symbol $\sqrt{\phantom{x}}$ to represent a square root.

$$\sqrt{4} = 2 \qquad \sqrt{25} = 5 \qquad \sqrt{81} = 9$$

You need to be able to **remember** the square numbers up to 15 × 15 and the corresponding square roots.

## Cubes and cube roots

When a number is multiplied by itself then multiplied by itself again, the answer is a **cube number**. You can write cube numbers using index notation.

| Multiplication | Index notation | Cube number |
|---|---|---|
| 2 × 2 × 2 | $2^3$ | 8 |
| 3 × 3 × 3 | $3^3$ | 27 |
| 5 × 5 × 5 | $5^3$ | 125 |
| 10 × 10 × 10 | $10^3$ | 1000 |

Cube numbers are the volumes of cubes with whole number side lengths.
There is more about volume on page 84.

27 cm³
3 cm
3 cm
3 cm

**Cube roots** are the opposite of cubes. You use the symbol $\sqrt[3]{\phantom{x}}$ to represent a cube root.

$$\sqrt[3]{8} = 2 \qquad \sqrt[3]{27} = 3 \qquad \sqrt[3]{1000} = 10$$

You need to be able to **remember** the cubes of 2, 3, 4, 5 and 10, and the corresponding cube roots.

## Worked example

Pavel says
'A square number cannot be a prime number.'
Is Pavel correct? Give a reason to support your answer. **(2 marks)**

Yes because a square number must have a factor other than 1 and itself. The square root of a square number is always one of its factors, so it can't be a prime number.

## Problem solved!

You are asked to give a **reason** to support your answer. Start your answer with: 'Yes because...' or 'No because...'. If you're not sure you could **try some numbers** to see what is going on:
Factors of 9 are 1, 3, 9, so 9 is **not prime**.
Factors of 49 are 1, 7, 49, so 49 is **not prime**.
Revise factors and prime numbers on page 11.

You will need to use problem-solving skills throughout your exam – **be prepared!**

## Now try this

 **1** Work out
  (a) $6^2$ **(1 mark)**
  (b) $\sqrt[3]{125}$ **(1 mark)**
  (c) $4^3$ **(1 mark)**
  (d) $\sqrt{121}$ **(1 mark)**

 You should know these without using your calculator.

 **2** Amy says
'The product of two square numbers is always an even number.'
Is she correct? Give an example to support your answer. **(1 mark)**

# Indices

Indices include square roots, cube roots and powers. You can use the **index laws** to simplify powers.

 $a^m \times a^n = a^{m+n}$

$4^3 \times 4^7 = 4^{3+7} = 4^{10}$

 $\dfrac{a^m}{a^n} = a^{m-n}$

$12^8 \div 12^3 = 12^{8-3} = 12^5$

 $(a^m)^n = a^{mn}$

$(7^3)^5 = 7^{3 \times 5} = 7^{15}$

 **Negative powers**

$a^{-n} = \dfrac{1}{a^n}$

$5^{-2} = \dfrac{1}{5^2} = \dfrac{1}{25}$

**Be careful!**
A **negative** power can still have a **positive** answer.

 **Reciprocals**

$a^{-1} = \dfrac{1}{a}$

This means that $a^{-1}$ is the **reciprocal** of $a$.
You can find the reciprocal of a fraction by turning it upside down.

$\left(\dfrac{5}{9}\right)^{-1} = \dfrac{9}{5}$

 **Powers of fractions**

$\left(\dfrac{a}{b}\right)^n = \dfrac{a^n}{b^n}$

$\left(\dfrac{3}{10}\right)^2 = \dfrac{3^2}{10^2} = \dfrac{9}{100}$

 **Combining rules**

You can apply the rules one at a time.

$\left(\dfrac{a}{b}\right)^{-n} = \left(\dfrac{b}{a}\right)^n = \dfrac{b^n}{a^n}$

$\left(\dfrac{2}{3}\right)^{-3} = \left(\dfrac{3}{2}\right)^3 = \dfrac{3^3}{2^3} = \dfrac{27}{8}$

## Powers of 0 and 1

Anything raised to the power 0 is equal to 1.

$6^0 = 1$    $1^0 = 1$    $7223^0 = 1$    $(-5)^0 = 1$

Anything raised to the power 1 is equal to itself.

$8^1 = 8$    $499^1 = 499$    $(-3)^1 = -3$

## Worked example

(a) Write $6 \times 6 \times 6 \times 6 \times 6$ as a single power of 6. **(1 mark)**

$6 \times 6 \times 6 \times 6 \times 6 = 6^5$

(b) Simplify $\dfrac{3^7 \times 3}{3^4}$ fully, leaving your answer in index form. **(2 marks)**

$\dfrac{3^7 \times 3}{3^4} = \dfrac{3^8}{3^4} = 3^4$

3 is the same as $3^1$. For part (b), use the rule $a^m \times a^n = a^{m+n}$ to simplify the numerator, then use $\dfrac{a^m}{a^n} = a^{m-n}$ to simplify the fraction. Remember to write down both steps of your working and give your answer as a power.

## Now try this

**1** Simplify, leaving your answers as a power of 5.

(a) $7^2 \times 7^8$ **(1 mark)**

(b) $7^9 \div 7^2$ **(1 mark)**

(c) $\dfrac{7^4 \times 7^5}{7^6}$ **(2 marks)**

(d) $(7^5)^3$ **(1 mark)**

**2** (a) Write $\dfrac{5 \times 5^8}{5^4}$ as a single power of 7. **(2 marks)**

(b) Work out $\dfrac{4 \times 4^{10}}{4^5 \times 4^3}$ **(2 marks)**

 **3** Write down the value of

(a) $4^0$ **(1 mark)**

(b) $4^{-1}$ **(1 mark)**

(c) $4^{-2}$ **(1 mark)**

(d) $\left(\dfrac{2}{5}\right)^2$ **(1 mark)**

(e) $\left(\dfrac{3}{4}\right)^{-3}$ **(1 mark)**

Worked solution video

# Estimation

You can estimate the answer to a calculation by rounding each number to **1 significant figure**, and then doing the calculation. You can use this method to check your answers, or to estimate calculations on your **non-calculator paper**. Here are two examples:

**1**  $4.32 \times 18.09 \approx 4 \times 20 = 80$
The answer is approximately equal to 80.

**2**  $327^2 \approx 300^2 = 3^2 \times 100^2 = 90000$
The answer is approximately equal to 90000.

$\approx$ means 'is approximately equal to'

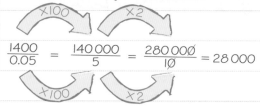

## Decimal division trick

You might have to divide by a decimal on your non-calculator paper. If you multiply both numbers in a division by the same amount the answer stays the same.

$$\frac{1400}{0.05} = \frac{140\,000}{5} = \frac{280\,000}{10} = 28\,000$$

## Worked example

Work out an estimate for

(a) $\dfrac{4.31 \times 278}{0.487}$    **(2 marks)**

$$\frac{4.31 \times 278}{0.487} \approx \frac{4 \times 300}{0.5} = \frac{1200}{0.5} = 2400$$

> Round all the numbers to **1 significant figure**. Then **write out** the calculation with the rounded values before calculating your estimate.

(b) $37.4^3$    **(2 marks)**

$$37.4^3 \approx 40^3 = 4^3 \times 10^3 = 64 \times 1000 = 64\,000$$

> You can use the laws of indices to work out $40^3$ without a calculator.
> $(ab)^n = a^n \times b^n$
> so $40^3 = (4 \times 10)^3 = 4^3 \times 10^3$

## Problem solved!

If you need to use formulae for spheres and cones they will be given with the question:

> Surface area of sphere $= 4\pi r^2$
> Volume of sphere $= \frac{4}{3}\pi r^3$
> where $r$ is the radius.

On your non-calculator paper you can use $\pi = 3.142$ then round to 1 s.f. to make your estimate.

> You will need to use problem-solving skills throughout your exam – **be prepared!**

## Worked example

A spherical ball-bearing has a radius of 2.35 cm.

(a) Work out an estimate for its surface area in square centimetres.    **(2 marks)**

$4\pi r^2 = 4 \times 3.142 \times 2.35^2$
$\approx 4 \times 3 \times 2^2 = 48 \text{ cm}^2$

(b) Is your answer to part (a) an overestimate or an underestimate? Give a reason for your answer.    **(1 mark)**

$3 < 3.142$ and $2 < 2.35$
so the answer is an underestimate.

## Now try this

**1** Use approximations to work out an estimate for

$$\frac{82 \times 285}{64 \times 35}$$

You **must** show your working.    **(2 marks)**

**2** A scientist models a raindrop as a sphere with radius 3.2 mm.

Volume of a sphere $= \frac{4}{3}\pi r^3$

(a) Work out an estimate for the volume of the raindrop.    **(2 marks)**

(b) Is your answer to part (a) an overestimate or an underestimate?
Give a reason for your answer.    **(1 mark)**

# Factors, multiples and primes

## Factors and multiples

The **factors** of a number are any whole numbers that divide into it exactly.

I and the number itself are both factors of any number.
The factors of 12 are 1, 2, 3, 4, 6 and 12.

Factors come in pairs. Each pair is a multiplication with the number as its answer.
The factor pairs of 12 are 1 × 12, 2 × 6 and 3 × 4.

A common factor is a number that is a factor of two or more numbers.
2 is a common factor of 6 and 12.

The **multiples** of a number are all the numbers in its times table.
The multiples of 7 are 7, 14, 21, 28, 35, ...

A common multiple is a number that is a multiple of two or more numbers.
12 is a common multiple of 6 and 4.

## Primes

A **prime number** has exactly two factors. It can only be divided by 1 and by itself.

The first ten prime numbers are
2, 3, 5, 7, 11, 13, 17, 19, 23, 29.
1 is not a prime number. It has only 1 factor.

### Factor trees

You can use a factor tree to find prime factors.
1. Choose a factor pair of the number.
2. Circle the prime factors as you go along.
3. Continue until every branch ends with a prime number.
4. At the end write down **all** the circled numbers, putting in multiplication signs.

---

## Worked example

(a) Which number is a multiple of 9?
Circle your answer. **(1 mark)**
16  8  30  43  6  ⑳27  12

(b) Which number is a prime number?
Circle your answer. **(1 mark)**
16  8  30  ㊸43  6  27  12

(c) Which number has eight factors?
Circle your answer. **(1 mark)**
16  8  ㉚30  43  6  27  12

The factors of 30 are:
1, 2, 3, 5, 6, 10, 15, 30

## Worked example

Write 90 as the product of prime factors.
Give your answer in index form. **(3 marks)**

$90 = 5 \times 2 \times 3 \times 3$
$= 5 \times 2 \times 3^2$

Use a factor tree to find all the prime factors. 3 appears twice in the factor tree, so you have to write $3^2$.

**Check it!**
$5 \times 2 \times 3^2 = 10 \times 9 = 90$ ✓

---

## Now try this

1 (a) Which number is a factor of 45?
Circle your answer. **(1 mark)**
2  9  18  20  24  30  36

(b) Which number is one less than a multiple of 7? Circle your answer. **(1 mark)**
2  9  18  20  24  30  36

(c) Which number is a multiple of 6 and a factor of 48? Circle your answer. **(1 mark)**
2  9  18  20  24  30  36

2 Write 280 as the product of its prime factors.
Give your answer in index form.
**(3 marks)**

Make sure you can recognise all the prime numbers below 50.

11

# HCF and LCM

The **highest common factor (HCF)** of two numbers is the highest number that is a factor of both numbers.

The **lowest common multiple (LCM)** of two numbers is the lowest number that is a multiple of both numbers.

For a reminder about prime factors and drawing a factor tree, have a look back at page 11.
Don't forget to write out the product of primes at the end.
**Check it!**
$2 \times 2 \times 3 \times 3 \times 3 = 108$ ✓

You can find the HCF and LCM by writing the products of prime factors in a **Venn diagram**. Use the powers to tell you how many times each prime factor occurs. Put the **common factors** in the intersection of the two ovals.
• HCF = product of all the prime factors in the intersection.
• LCM = product of all the prime factors in the Venn diagram.
There is more on Venn diagrams on page 131.

The question asks for the HCF and LCM as a product of powers of prime factors. You don't need to work out the actual values, but you do need to make sure you use **index notation** (powers).
For a reminder about this look at page 9.

## Worked example

(a) Express 108 as a product of powers of its prime factors. **(3 marks)**

$108 = 2 \times 2 \times 3 \times 3 \times 3$
$= 2^2 \times 3^3$

(b) $240 = 2^4 \times 3 \times 5$
Find, as a product of powers of its prime factors
(i) the highest common factor (HCF) of 108 and 240. **(1 mark)**

Factors of 108       Factors of 240

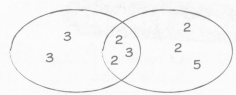

$HCF = 2 \times 2 \times 3 = 2^2 \times 3$
(ii) the lowest common multiple (LCM) of 108 and 240. **(1 mark)**
$LCM = 2 \times 2 \times 2 \times 2 \times 3 \times 3 \times 3 \times 5$
$= 2^4 \times 3^3 \times 5$

## Worked example

$n$ is a number.
10 is the highest common factor of 20 and $n$.
Which **two** numbers are possible values for $n$?
Circle the correct answers. **(2 marks)**

8   ⑩   25   ㉚   40   55

You can use prime factors to answer this question.
$20 = 2 \times 2 \times 5$
$10 = 2 \times 5$
Any number which contains the prime factors 2 and 5 exactly once will work.

## Now try this

1 (a) Write 132 and 110 as products of powers of their prime factors. **(4 marks)**
   (b) Find the HCF of 132 and 110. **(1 mark)**
   (c) Find the LCM of 132 and 110. **(1 mark)**

Worked solution video

2 Given that $X = 3^2 \times 5 \times 7^4$ and $Y = 2 \times 3^3 \times 5^2$ write down, as a product of powers of its prime factors,
   (a) the HCF of $X$ and $Y$. **(1 mark)**
   (b) the LCM of $X$ and $Y$. **(1 mark)**

Expand the index notation, then draw a Venn diagram showing the prime factors of each number.

# Fractions

You need to be able to work confidently with fractions, with or without a calculator.

## 1 Dividing objects

You can use fractions to divide an object into parts.

The top number is called the **numerator**.

$\frac{2}{3}$ of this rectangle is shaded.

The bottom number is called the **denominator**.

## 2 Equivalent fractions

Different fractions can describe the same amount.

 $\frac{1}{2} = \frac{2}{4}$

$\frac{1}{2}$ and $\frac{2}{4}$ are called equivalent fractions.

You can find equivalent fractions by multiplying or dividing the numerator and denominator by the same number.

## 3 Cancelling fractions

To **cancel** or **reduce** a fraction you divide the top and bottom by the same number.

When you can't cancel the fraction any further it is in its **simplest form**.

## 4 Finding a fraction of an amount

| Divide the amount by the denominator |
| --- |

↓

| Multiply by the numerator |
| --- |

Work out $\frac{3}{10}$ of 200 kg:

$200\,kg \div 10 = 20\,kg$

$20\,kg \times 3 = 60\,kg$

To see how to convert between fractions and decimals see page 57.

---

## Worked example

(a) Circle the fraction that is **not** equivalent to $\frac{5}{6}$     **(1 mark)**

$\frac{10}{12}$   ⟨$\frac{16}{18}$⟩   $\frac{20}{24}$   $\frac{35}{42}$

(b) Work out $\frac{2}{5}$ of £240

$240 \div 5 = 48$

$48 \times 2 = 96$

$\frac{2}{5}$ of £240 is £96     **(2 marks)**

You should simplify **all** the fractions as much as possible:

**Check it!**

Only **one** fraction is not equivalent to $\frac{5}{6}$. Check that this matches your working.

---

## Now try this

Plan your strategy before you start. Work out how much money he makes in total, compare this with £8 then write a conclusion.

1. David buys 80 chocolates for £20.
   He sells $\frac{3}{4}$ of the chocolates for 40p each.
   He then sells the remaining chocolates for 20p each.
   Work out the total profit that David makes.    **(4 marks)**

2. Sandeep buys some pencils to sell at the school fete.
   He buys 40 pencils for £8.
   He sells a quarter of the pencils for 25p each and sells half of the pencils for 30p each.
   The remaining pencils are broken and are thrown away.
   Does Sandeep make a profit or loss?
   You must show your working.    **(4 marks)**

# Operations on fractions

Make sure you can add, subtract, multiply and divide fractions **without** a calculator.

## ① Adding or subtracting

| Write both fractions as equivalent fractions with the same denominator |
| :---: |
| ↓ |
| Add or subtract the numerators |
| ↓ |
| Do not change the denominator |

### Worked example

  Work out
(a) $\frac{1}{5} + \frac{3}{10}$     (b) $\frac{8}{9} - \frac{1}{6}$

   **(2 marks)**          **(2 marks)**

$= \frac{2}{10} + \frac{3}{10}$       $= \frac{16}{18} - \frac{3}{18}$

$= \frac{5}{10} = \frac{1}{2}$        $= \frac{13}{18}$

> 18 is the lowest common multiple (LCM) of 9 and 6. This is the easiest common denominator to use.
> For a reminder about LCMs see page 12.

## ② Multiplying

| Write any whole numbers on their own as fractions with denominator 1 |
| :---: |
| ↓ |
| Multiply the numerators and multiply the denominators |

### Worked example

Work out
(a) $\frac{2}{3} \times \frac{7}{10}$ **(2 marks)**    (b) $3 \times \frac{2}{11}$ **(2 marks)**

$= \frac{2 \times 7}{3 \times 10}$          $= \frac{3}{1} \times \frac{2}{11}$

$= \frac{14}{30} = \frac{7}{15}$      $= \frac{6}{11}$

## ③ Dividing

| Write any whole numbers on their own as fractions with denominator 1 |
| :---: |
| ↓ |
| Turn the second fraction 'upside down' |
| ↓ |
| Change ÷ to × |
| ↓ |
| Multiply the numerators and multiply the denominators |

### Worked example

Work out
(a) $\frac{2}{5} \div \frac{3}{4}$ **(2 marks)**    (b) $6 \div \frac{2}{3}$ **(2 marks)**

$= \frac{2}{5} \times \frac{4}{3}$        $= \frac{6}{1} \div \frac{2}{3}$

$= \frac{8}{15}$            $= \frac{6}{1} \times \frac{3}{2}$

> Change $\frac{3}{4}$ into $\frac{4}{3}$ and change ÷ to ×.

                       $= \frac{18}{2}$

                       $= 9$

## Watch out!

1. You do not have to cancel your final answer unless the question asks you to 'give your answer in its simplest form'.

2. You can compare and order fractions by using equivalent fractions with the same denominator.

See page 13 for a reminder about equivalent fractions and simplest form.

### Now try this

 **1** Work out
(a) $\frac{5}{9} + \frac{1}{3}$    **(2 marks)**
(b) $\frac{3}{10} \div \frac{8}{15}$    **(2 marks)**

  **2** Work out
(a) $\frac{7}{8} - \frac{1}{2}$ **(2 marks)**    (b) $\frac{8}{11} - \frac{3}{5}$ **(2 marks)**

 **3** Work out
(a) $\frac{5}{9} \times \frac{3}{10}$ **(2 marks)**    (b) $\frac{8}{15} \div \frac{4}{7}$ **(3 marks)**

Worked solution video

# Mixed numbers

**Mixed numbers** have a whole number part and a fraction part.

$3\frac{1}{4}$   This mixed number is the same as $3 + \frac{1}{4}$

**Improper fractions** have a numerator larger than their denominator.

$\frac{5}{2}$, $\frac{21}{5}$ and $\frac{4}{3}$ are all improper fractions.

## Converting between mixed numbers and improper fractions

To convert a mixed number into an improper fraction you...

Multiply this ...

$3\frac{1}{4} = \frac{3 \times 4 + 1}{4} = \frac{13}{4}$

...by this...    ...add it to this.    Keep the same
$3 \times 4 = 12$    $12 + 1 = 13$    denominator.

To convert an improper fraction into a mixed number you...

Divide this ...

$\frac{23}{5} = 23 \div 5 = 4\frac{3}{5}$

Write the remainder as the numerator.

...by this.    Keep the same denominator.

> **Golden rule**
>
> You need to write mixed numbers as improper fractions before you do any calculations.

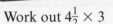

## Worked example

Work out $4\frac{1}{2} \times 3$    **(2 marks)**

$= \frac{9}{2} \times \frac{3}{1}$      $4\frac{1}{2} = \frac{4 \times 2 + 1}{2} = \frac{9}{2}$

$= \frac{27}{2}$

$= 13\frac{1}{2}$

## Worked example

Work out $3\frac{2}{5} - 1\frac{1}{2}$    **(3 marks)**

$= \frac{17}{5} - \frac{3}{2}$

$= \frac{34}{10} - \frac{15}{10} = \frac{19}{10}$

$= 1\frac{9}{10}$

## Worked example

Kumar has $2\frac{1}{2}$ pizzas. He wants to divide them equally between 3 people.
What fraction of a pizza does each person receive?    **(2 marks)**

$2\frac{1}{2} \div 3 = \frac{5}{2} \div \frac{3}{1}$

$= \frac{5}{2} \times \frac{1}{3}$

$= \frac{5}{6}$

Each person receives $\frac{5}{6}$ of a pizza.

Follow these steps to add or subtract mixed numbers:

- Write both numbers as improper fractions.
- Write both fractions as equivalent fractions with the same denominator.
- Add or subtract the numerators.
- If your answer is an improper fraction, convert it into a mixed number.

## Now try this

**1** Work out
   (a) $3\frac{1}{3} \times 1\frac{3}{4}$    **(3 marks)**
   (b) $4\frac{1}{5} \div 1\frac{7}{8}$    **(3 marks)**

Write both numbers as improper fractions before you do the calculation. Remember you can't use a calculator for these questions.

**2** A cup holds $\frac{1}{3}$ litre of water.
   How many **full** cups can be filled from a $2\frac{1}{2}$ litre jug of water?    **(3 marks)**

# Calculator and number skills

These calculator keys are really useful:

$x^2$    Square a number.

$(-)$    Enter a negative number.

$x^3$    Cube a number.

$\sqrt{\Box}$    Find the square root of a number.

$x^{-1}$    Find the reciprocal of a number.

$\sqrt[3]{\Box}$    Find the cube root of a number. You might need to press the shift key first.

Ans    Use your previous answer in a calculation.

S⇔D    Change the answer from a fraction or surd to a decimal. Not all calculators have this key.

## Order of operations

You need to use the correct **priority of operations** when doing a calculation.

**B**rackets

**I**ndices

**D**ivision

**M**ultiplication

**A**ddition

**S**ubtraction

$(10 - 7) + 4 \times 3^2$
$= 3 + 4 \times 3^2$
$= 3 + 4 \times 9$
$= 3 + 36$
$= 39$

## Reciprocals

To find the **reciprocal** of a number you write it as a fraction then turn the fraction upside down.

$7 = \frac{7}{1} \rightarrow \frac{1}{7}$    The reciprocal of 7 is $\frac{1}{7}$

$\frac{3}{4} \rightarrow \frac{4}{3}$    The reciprocal of $\frac{3}{4}$ is $\frac{4}{3}$

You can use the $x^{-1}$ key on your calculator to find reciprocals.

## Calculating with fractions

You can enter fractions on your calculator using the ▤ key and the arrows.

For example, to work out $\frac{3}{8}$ of 52:

$\frac{3}{8} \times 52$

$\frac{39}{2}$

If you want to convert an answer on your calculator display from a fraction to a decimal you can use the S⇔D key.

## Worked example

(a) Find the value of $\sqrt{6 - 0.71}$    **(1 mark)**

$6 - 0.71 = 5.29$
$\sqrt{5.29} = 2.3$

(b) Find the value of $3.1^4$    **(1 mark)**

$92.3521$

(a) The square root acts like a **bracket** so you have to work out everything inside first. If you work this out in one go using the $\sqrt{\Box}$ key on your calculator, check that **all** of the calculation is underneath the square root sign.

(b) Use the $x^{\Box}$ key to work out powers of 4 and higher. Write down **all the digits** from your calculator display.

## Now try this

Try doing these **without a calculator**. Make sure you use the correct priority of operations.

**1** Work out
(a) $10 - 3 \times 2$    **(1 mark)**
(b) $84 \div (6 + 6)$    **(1 mark)**
(c) $8 + 3^2$    **(1 mark)**
(d) $(9 - 4)^3$    **(1 mark)**

**2** Work out
(a) $\sqrt{244 + 717}$    **(1 mark)**
(b) $\sqrt[3]{3.1 - 0.356}$    **(1 mark)**

Use a calculator and give your answers as decimal numbers. You might need to use the S⇔D key to convert a fraction answer to a decimal.

**Worked solution video**

# Standard form 1

Numbers in standard form are written as the product two parts.

$$7.3 \times 10^{-6}$$

This part is a number greater than or equal to 1 and less than 10

This part is an integer power of 10

You can use standard form to write very large or very small numbers.

$$920\,000 = 9.2 \times 10^5$$

Numbers greater than 10 have a positive power of 10

$$0.00703 = 7.03 \times 10^{-3}$$

Numbers less than 1 have a negative power of 10

## Counting decimal places

You can count decimal places to convert between numbers in standard form and ordinary numbers.

3 jumps

$$7\,9\,0\,0 = 7.9 \times 10^3$$

7900 > 10 so the power is positive

4 jumps

$$0.0\,0\,0\,3\,5 = 3.5 \times 10^{-4}$$

0.00035 < 1 so the power is negative

**Be careful!**
Don't just count zeros to work out the power.

---

## Worked example

(a) Write 1 630 000 in standard form.  **(1 mark)**
1.63 × 10⁶

(b) Write 4.2 × 10⁻³ as an ordinary number.  **(1 mark)**
0.0042

Count the number of decimal places you need to move to get a number between 1 and 10. 1 630 000 is bigger than 10 so the power will be positive.

---

## Using a calculator

You can enter numbers in standard form using the ×10ˣ key.
To enter 3.7 × 10⁻⁶ press

If you are using a calculator with numbers in standard form it is a good idea to put brackets around each number.

Your calculator might give the answer as a normal number. You need to convert it into standard form.

## Worked example

$A$ and $B$ are standard form numbers.
$A = 1.9 \times 10^4$    $B = 4.2 \times 10^5$
Calculate, giving your answers in standard form:

(a) $A + B$  **(1 mark)**
(1.9 × 10⁴) + (4.2 × 10⁵) = 439 000
= 4.39 × 10⁵

(b) $B \div 6$  **(1 mark)**
(4.2 × 10⁵) ÷ 6 = 70000 = 7 × 10⁴

---

## Now try this

1  (a) Write 0.00028 in standard form.  **(1 mark)**
    (b) Write 3.91 × 10⁵ as an ordinary number.  **(1 mark)**

2  $p$ and $q$ are standard form numbers.
   $p = 3.5 \times 10^6$    $q = 7.9 \times 10^5$
   Calculate, giving your answers in standard form:
   (a) $p - q$  **(1 mark)**
   (b) $5p$  **(1 mark)**

# Standard form 2

This page shows **non-calculator** methods for doing calculations with standard form.

## 1 Multiplying numbers in standard form

| Rearrange so powers of 10 are together |
| --- |

$$(3 \times 10^3) \times (5 \times 10^6)$$

| Multiply the number parts |
| --- |

$$= (3 \times 5) \times (10^3 \times 10^6)$$
$$= 15 \times 10^9$$

| Add the powers |
| --- |

$$a^m \times a^n = a^{m+n}$$

| Rewrite your answer in standard form if necessary |
| --- |

$$= (1.5 \times 10^1) \times 10^9$$
$$= 1.5 \times 10^{10}$$

## 2 Dividing numbers in standard form

| Rearrange so powers of 10 are together |
| --- |

$$(1.2 \times 10^8) \div (2 \times 10^4)$$

| Divide the number parts |
| --- |

$$= (1.2 \div 2) \times (10^8 \div 10^4)$$
$$= 0.6 \times 10^4$$

| Subtract the powers |
| --- |

$$a^m \div a^n = a^{m-n}$$

| Rewrite your answer in standard form if necessary |
| --- |

$$= 6 \times 10^{-1} \times 10^4$$
$$= 6 \times 10^3$$

## 3 Adding and subtracting numbers in standard form

| Write both numbers as ordinary numbers |
| --- |

$$3.9 \times 10^5 + 2.5 \times 10^4$$
$$= 390\,000 + 25\,000$$

| Add or subtract normally |
| --- |

$$\begin{array}{r} 390\,000 \\ +\ 25\,000 \\ \hline 415\,000 \end{array}$$

| Convert the answer back into standard form |
| --- |

$$= 415\,000$$
$$= 4.15 \times 10^5$$

### Worked example

$A$ and $B$ are standard form numbers.
$A = 8.2 \times 10^6$   $B = 2 \times 10^4$
Calculate $A \times B$ giving your answer in standard form. **(2 marks)**

$$(8.2 \times 10^6) \times (2 \times 10^4)$$
$$= (8.2 \times 2) \times (10^6 \times 10^4)$$
$$= 16.4 \times 10^{10}$$
$$= 1.64 \times 10^1 \times 10^{10}$$
$$= 1.64 \times 10^{11}$$

### Problem solved!

If you're not sure which operation to use, try using **simpler numbers**:

An egg weighs 80 g and a box of eggs weighs 960 g. Calculate the number of eggs in the box.

You would divide the total weight by the weight of each egg. So for this example, divide the mass of the gold bar by the mass of one atom.

You will need to use problem-solving skills throughout your exam – **be prepared!**

### Worked example

A single atom of gold has a mass of $3 \times 10^{-22}$ g.
A gold bar has a mass of 1.2 kg.
Calculate the number of atoms of gold in the bar. **(3 marks)**

$$1.2\,\text{kg} = 1200\,\text{g} = 1.2 \times 10^3\,\text{g}$$
$$\frac{1.2 \times 10^3}{3 \times 10^{-22}} = \frac{1.2}{3} \times \frac{10^3}{10^{-22}}$$
$$= 0.4 \times 10^{25}$$
$$= 4 \times 10^{-1} \times 10^{25}$$
$$= 4 \times 10^{24}$$

### Now try this

Worked solution video

**1** (a) Work out
$(3 \times 10^7) \times (9 \times 10^4)$.
Write your answer in standard form. **(2 marks)**

(b) Work out
$(6.4 \times 10^{-9}) \div (2 \times 10^{-3})$.
Write your answer in standard form. **(2 marks)**

# Counting strategies

You might need to make a list of possible combinations. You need to find a **systematic** way of doing this to make sure you have found every possible combination. You could use the three number cards on the right to make **six** different three-digit numbers. Here is one way of writing them out systematically.

Start by writing out the numbers that begin with 4.    Then write the numbers beginning with 5.    Finally, write the numbers that begin with 6.

## Worked example

James is making badges to sell. He can choose three different colours:
Red (R)   Green (G)   Blue (B)
He can also choose three shapes:
Circle (C)   Star (S)   Oval (O)
How many possible combinations of badge colour and shape can James make?    **(2 marks)**

| RC | GC | BC |
| RS | GS | BS |
| RO | GO | BO |

There are 9 possible combinations.

> You need to use a systematic method to make sure you have written down every possible combination. Start by writing down all the possible red badges. Then write down all the possible green badges and then all the possible blue badges.

> Make sure you **answer the question**. You need to show a systematic strategy **and** write the number of combinations.

## Problem solved!

You can save time in your exam by numbering a list of items like this. You can still clearly **show your strategy** using the numbers, and you don't have to write as much.
Read the question carefully. Each team must play the other teams only once, so be careful not to count each match more than once.

> You will need to use problem-solving skills throughout your exam – **be prepared!**

## Worked example

A school hockey league contains five teams:
Benfield ①   Gosforth ②   Kenton ③
St Cuthbert's ④   Sacred Heart ⑤
Each team must play each of the other teams once.
How many matches will be played in total?    **(2 marks)**

1 vs 2, 1 vs 3, 1 vs 4, 1 vs 5
2 vs 3, 2 vs 4, 2 vs 5
3 vs 4, 3 vs 5
4 vs 5     There will be 10 matches in total.

## Now try this

 **1** Seth has three tickets to go to the cinema. He chooses two friends to go with him. He chooses one friend from Ali, Ben and Carl, and one friend from Yvonne and Zara.
List all the possible combinations of friends that Seth could choose.    **(2 marks)**

 **2** A local darts tournament has six clubs competing for a place in the national finals: Compton, Penn, Goldthorn, Lanesfield, Tettenhall, Wightwick.
Over a four-week period each club must play all the other clubs once each.
How many darts matches will be played in total during the tournament?    **(2 marks)**

# Problem-solving practice 1

About half of the questions in your Foundation GCSE exam will require you to **problem-solve, reason, interpret** or **communicate** mathematically. If you come across a tricky or unfamiliar question in your exam you can try some of these strategies:

- ☑ Sketch a diagram to see what is going on.
- ☑ Try the problem with smaller or easier numbers.
- ☑ Plan your strategy before you start.
- ☑ Write down any formulae you might be able to use.
- ☑ Use *x* or *n* to represent an unknown value.

**1** Wilfred has these coins.

Nisha has these coins.

Wilfred gives Nisha one coin. Wilfred now has twice as much money as Nisha.

What value is the coin Wilfred gives to Nisha?

(2 marks)

## Place value page 1

In money questions you should decide whether you are going to work in pounds or pence. Don't mix up your units.

Remember that if Wilfred gives a coin to Nisha his amount of money is **reduced** by that value and hers is **increased** by the same amount.

**TOP TIP**

With a question like this it's easy to check your answer. Write down the new amounts of money Wilfred and Nisha have and check that Wilfred's total is twice Nisha's.

**2** Liam is planning a trip for a group of 20 children and 5 adults.

They can go to either the theatre or the zoo.

If they go to the theatre, they will go by train.

If they go to the zoo, they will go by coach.

Liam has information about the costs.

| Theatre ticket prices | Return train fares |
|---|---|
| Stalls: £22 | Adults: £11.50 |
| Circle: £15 | Child: £5.75 |

| Zoo admission | Coach hire |
|---|---|
| Adult: £18 | 20 Seats: £190 |
| Child: £12 | 30 Seats: £240 |
| | 40 Seats: £300 |

What is the lowest possible total cost of the trip? You must show all your working.

(5 marks)

## Calculator and number skills page 16

Work out the total cost of each trip. Remember to choose the cheapest ticket price for the theatre tickets, and write down all your working. You have to say which trip is the cheapest to complete your answer.

**TOP TIP**

Plan how you will lay out your answer. You need to show **what** you are working out at each stage, so write short headings to go with your workings. You could use these headings:

Cost of theatre trip
Cost of zoo trip
Conclusion

# Problem-solving practice 2

**3** The diagram shows three identical shapes.

$\frac{3}{4}$ of shape A is shaded and $\frac{3}{5}$ of shape C is shaded.

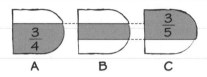

A      B      C

What fraction of shape B is shaded?

(3 marks)

## Fractions page 13

If the whole of each shape is 1, then you can start by working out the fraction that represents the **white** areas on shapes A and C. You can then subtract these from 1 to find the shaded area on shape B.

**TOP TIP**

You have to **show your strategy**. This means you need to show all your working so it is clear how you have tackled the problem.

---

**4** Susan has 2 dogs.
Each dog is fed $\frac{3}{8}$ kg of dog food each day.

Susan buys dog food in bags.

Each bag weighs 14 kg.

For how many days can Susan feed the 2 dogs from 1 bag of dog food?

You must show **all** your working.   (5 marks)

## Fractions page 13

There are lots of steps in this question so make sure you keep track of your working.

**TOP TIP**

If your working is very untidy or hard to follow, then re-write it clearly and cross out your original working.

---

**5** The diagram shows two types of plastic building block.

A          B

←24 mm→    ←——32 mm——→

Block A is 24 mm long.

Block B is 32 mm long.

Jeremy joins some type A blocks together to make a straight row.

He then joins some type B blocks together to make a straight row of the same length.

Write down the shortest possible length of this row.   (4 marks)

Worked solution video

## HCF and LCM page 12

You have to use a **whole number** of building blocks in each row, so the length of each row has to be a multiple of the length of one block. The answer will be the smallest number that is a multiple of 24 and a multiple of 32. This is the **lowest common multiple** of 24 and 32. You can't get all the marks just by writing down the answer. You need to show, clearly and neatly, how you got your answer.

**TOP TIP**

If you're not sure how to start you could draw a sketch or a scale drawing:

This might help you see that the lengths are multiples of 24 and 32.

# Collecting like terms

## Expressions, equations and formulae

In algebra you use letters to represent unknown numbers.

$$4x + 3y - z$$

This is an **expression**. It does not have an = sign. The parts which are separated by + or − signs are called **terms**.

$$3n - 1 = 17$$

This is an **equation**. This equation only has one letter in it. You can solve an equation to find the value of the letter.

$$A = \frac{1}{2}bh$$

This is a **formula**. You can use it to calculate one value if you know the other values. You can't solve a formula.

## Simplifying expressions

You can simplify expressions that contain + and − by **collecting like terms**.

Like terms contain the same letters.

$h + h + h = \boxed{3h}$ — This means '3 lots of $h$' or $3 \times h$.

$5x - 2x = \boxed{3x}$ — '5 lots of $x$' minus '2 lots of $x$' equals '3 lots of $x$'.

$2p + 3q - 5p + q = 2p - 5p + 3q + q$

$= -3p + 4q$

$2p$ and $-5p$ are like terms.

$2p - 5p = -3p$

Look at page 23 for more on simplifying expressions.

### Golden rules

1. Each term includes the sign (+ or −) in front of it.

2. $x$ means '1 lot of $x$'. You don't need to write $1x$

3. Like terms contain exactly the same combinations of letters with the same powers.

Like terms
$xy$     $-3xy$
$+10xy$   $-xy$

NOT like terms
$3a$     $+a^2$
$-2ab$   $-5a^2bc$

---

**Worked example**

(a) Simplify $x + x + x + x$     **(1 mark)**

$4x$

(b) Simplify $3n + n + 5n$     **(1 mark)**

$9n$

Remember $n$ means '1 lot of $n$'. Here there are $3 + 1 + 5 = 9$ lots of $n$ in total. You write this as $9n$.

---

**Worked example**

Simplify $3m + 6b - 2m + b$     **(2 marks)**

$3m + 6b - 2m + b = 3m - 2m + 6b + b$

$= m + 7b$

Group the $m$ terms together and group the $b$ terms together. Remember that each term includes the sign in front of it.

---

**Now try this**

Worked solution video

1 (a) Simplify
   $x + x + x + x + x + x$     **(1 mark)**
  (b) Simplify $5q + 3q$     **(1 mark)**
  (c) Simplify $11t - t$     **(1 mark)**
  (d) Simplify $6n + n - 4n$     **(1 mark)**

2 | root   formula   factor   term |

Choose a word from the list above to make this sentence correct:

$8x$ is a _____ in $3y + 8x + 5$     **(1 mark)**

3 Simplify $8p + 2q + p - 5q$     **(2 marks)**

# Simplifying expressions

You'll need to be able to **simplify** expressions that contain $\times$ and $\div$ in your exam. Use these rules to help you.

 **Multiplying expressions**

1. Multiply any number parts first.
2. Then multiply the letters. Remember to use ☐$^2$ for letters which are multiplied twice or ☐$^3$ for letters which are multiplied three times.

$$10a \times 3a = 30a^2$$

$10 \times 3 = 30$　　　$a \times a = a^2$

$$3s \times 6t = 18st$$

$3 \times 6 = 18$　　　$s \times t = st$

**Dividing expressions**

1. Write the division as a fraction.
2. Cancel any number parts.
3. If the same letter appears on the top and bottom, you can cancel that as well.

$$8y \div 4 = \frac{^2\cancel{8}y}{\cancel{4}_1} = 2y$$　　$8 \div 4 = 2$

$$\frac{^4\cancel{36}a\cancel{b}}{_1\cancel{9}\cancel{b}} = 4a$$　　$36 \div 9 = 4$

$b$ appears on the top and the bottom, so cancel

---

## Multiplying with algebra

You can multiply letters in algebra by writing them next to each other.
$ab = a \times b$

You can use indices to describe a letter multiplied by itself.
$y \times y = y^2$　You say 'y squared'.

You can use indices to describe the same letter multiplied together three times.
$n \times n \times n = n^3$　You say 'n cubed'.

For a reminder about squares and cubes have a look at page 8.

---

## Worked example

(a) Simplify $a \times a \times a$　　**(1 mark)**
$a^3$

(b) Simplify $7x \times 2y$　　**(2 marks)**
$14xy$

(c) Simplify $10pq \div 2p$　　**(2 marks)**
$$\frac{^5\cancel{10}\cancel{p}q}{_1\cancel{2}\cancel{p}} = 5q$$

(b) Multiply the numbers first and then the letters.
$7 \times 2 = 14$　　　　$x \times y = xy$

(c) Write the division as a fraction. You can cancel the number parts by dividing top and bottom by 2.
$10 \div 2 = 5$ so write 5 on top of your fraction.
$p$ appears on the top and the bottom so you can cancel it. You are left with $5q$.

---

## Now try this

Simplify
(a) $n \times n \times n \times n \times n$　　**(1 mark)**
(b) $3r \times 2r \times r$　　**(2 marks)**
(c) $4y \times 8z$　　**(2 marks)**
(d) $20fg \div 5g$　　**(2 marks)**
(e) $28wk \div 4wk$　　**(2 marks)**

(d) $20fg \div 5g = \dfrac{20fg}{5g}$

Cancel any number parts then cancel any letters that are both on the top and bottom of the fraction.

# Algebraic indices

You revised indices with numbers on page 9. You can also use the index laws to simplify powers in algebraic expressions.

 $y^3$
- This part is called the **index**. The plural of index is indices.
- This part is called the **base**.

$n^5$  This means $n \times n \times n \times n \times n$.

---

## Index laws

You can use these three index laws to simplify powers and algebraic expressions.

 **1** To multiply powers of the same base, add the indices.

$$a^m \times a^n = a^{m+n}$$
$$x^4 \times x^3 = x^{4+3} = x^7$$

 **2** To divide powers of the same base, subtract the indices.

$$a^m \div a^n = \frac{a^m}{a^n} = a^{m-n}$$
$$\frac{m^8}{m^2} = m^{8-2} = m^6$$

 **3** To raise a power of a base to a further power, multiply the indices.

$$(a^m)^n = a^{mn}$$
$$(j^2)^4 = j^{2\times4} = j^8$$

 **4** Powers apply to everything inside the brackets.

$$(ab)^n = a^n \times b^n$$
$$(4b)^3 = 4^3 \times b^3 = 64b^3$$

### One at a time

When you are multiplying or dividing expressions with powers:

1.  Multiply or divide any number parts first.

2.  Use the index laws to work out the new power.

$7x \times 5x^6 = 35x^7$ ← $x \times x^6 = x^{1+6} = x^7$

$7 \times 5 = 35$

$12 \div 3 = 4$

$$\frac{12a^5}{3a^2} = 4a^3$$

$a^5 \div a^2 = a^{5-2} = a^3$

(a) Add the indices. $a^m \times a^n = a^{m+n}$
(b) Subtract the indices. $a^m \div a^n = a^{m-n}$
(c) Multiply the indices. $(a^m)^n = a^{mn}$

---

## Worked example

(a) Simplify $p^2 \times p^7$  **(1 mark)**
$p^2 \times p^7 = p^{2+7} = p^9$

(b) Simplify $m^8 \div m^3$  **(1 mark)**
$m^8 \div m^3 = m^{8-3} = m^5$

(c) Simplify $(a^6)^3$  **(1 mark)**
$(a^6)^3 = a^{6\times3} = a^{18}$

### Watch out!

You can only use the index laws when the bases are the same.

If there's no index then the number has a power of 1.

$n \times n^5 = n^{1+5} = n^6$
$x^8 \div x = x^{8-1} = x^7$

---

## Now try this

**1** (a) Simplify $m^5 \times m^4$  **(1 mark)**
  (b) Simplify $k^6 \times k$  **(1 mark)**

**2** (a) Simplify $h^{11} \div h^3$  **(1 mark)**
  (b) Simplify $\dfrac{t^{25}}{t^5}$  **(1 mark)**
  (c) Simplify $\dfrac{a^3 \times a^6}{a^4}$  **(1 mark)**

**3** (a) Simplify $(y^4)^2$  **(1 mark)**

  (b) Simplify $(5p^2)^3$  **(1 mark)**

You could do this in two different ways:
$(5p^2)^3 = 5^3 \times (p^2)^3$
$(5p^2)^3 = 5p^2 \times 5p^2 \times 5p^2$
Both ways will give you the same answer.

# Substitution

If you know the values of the letters in an algebraic expression, you can **substitute** them into the expression. This lets you work out the value of the expression.

$x = 7$ and $y = 2$ have been substituted into this expression.

$x$   +   $5y$ ———• $5y$ means $5 \times y$.
                When $y = 2$, $5y = 5 \times 2 = 10$

$x = 7$    $y = 2$

$7 + 5 \times 2 = 7 + 10$   Multiply before you
$= 17$   Add.

When $x = 7$ and $y = 2$ the value of $x + 5y$ is 17.

## Using BIDMAS

Remember to use the correct priority of operations when you are doing a calculation. BIDMAS gives the priority in which the operations should be carried out.

**B**rackets
**I**ndices
**D**ivision
**M**ultiplication
**A**ddition
**S**ubtraction

$2 \times 5^2 - (14 + 8)$
$= 2 \times 5^2 - 22$
$= 2 \times 25 - 22$
$= 50 - 22$
$= 28$

You should substitute all the values before doing any calculations.

## Worked example

Be extra careful when substituting a negative number. You can use brackets around the number to make sure you don't make a mistake.

(a) Work out the value of $5x + 1$ when $x = -3$
**(2 marks)**

$5 \times (-3) + 1 = -15 + 1$
$\qquad = -14$

(b) Work out the value of $5p^3$ when $p = 2$
**(2 marks)**

$5 \times 2^3 = 5 \times 8$
$\qquad = 40$

Remember **BIDMAS**. Indices comes before Multiplication.

Substitute all the values before starting your calculation.

(c) Work out the value of $3m + 4n$ when $m = 5$ and $n = -2$   **(2 marks)**

$3 \times 5 + 4 \times (-2) = 15 + (-8)$
$\qquad = 15 - 8$
$\qquad = 7$

(d) Work out the value of $2x(x - 1)$ when $x = 11$
**(2 marks)**

$2 \times 11 \times (11 - 1) = 2 \times 11 \times 10$
$\qquad = 22 \times 10$
$\qquad = 220$

You can multiply in any order.
$22 \times 10 = 2 \times 110$

## Now try this

Pick two too.

**1** Work out the value of $h + k - 6n$ when $h = 5$, $k = 10$ and $n = \frac{1}{2}$   **(2 marks)**

**2** Work out the value of $\dfrac{xw}{x + y}$ when $x = -8$, $w = 3$ and $y = 4$   **(3 marks)**

First work out what the top and bottom of the fraction come to, then work out the division.

**3** $x = 3$, $y = 6$ and $z = 10$
Work out the value of these expressions.
(a) $x + z$   **(1 mark)**
(b) $yz - x^2$   **(1 mark)**
(c) $x^3 - (z - y)$   **(1 mark)**

Worked solution video

# Formulae

A **formula** is a mathematical rule.

Formulae is the plural of formula.

The formula for the area of this triangle is:

Area $= \frac{1}{2} \times$ base $\times$ height

You can write this formula using algebra as:

$A = \frac{1}{2}bh$

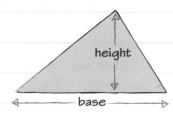

A formula lets you calculate one quantity when you know the others.

You need to **substitute** the values you know into the formula.

For more on substitution have a look at page 25.

---

## Worked example

You can use this formula to work out the cooking time in minutes for a turkey.

| Cooking time = Weight in kg $\times$ 30 + 45 |

Work out the cooking time for a turkey weighing 7 kg. **(3 marks)**

Cooking time $= 7 \times 30 + 45$

$\quad = 210 + 45$

$\quad = 255$

The cooking time is 255 minutes, or 4 hours and 15 minutes.

Substitute the weight into your formula before you do any calculations.

Remember to use **BIDMAS** for the correct order of operations. You **M**ultiply before you **A**dd.

For a reminder about substituting and BIDMAS have a look at page 25.

When you are giving an answer in minutes and it is larger than 60 minutes, you can give it in minutes or in hours and minutes.

$4 \times 60 = 240$ and $255 - 240 = 15$

So 255 minutes = 4 hours and 15 minutes.

---

Substitute the values for $u$ and $t$ into the formula. Using **BIDMAS**:

1. Do the **I**ndex (power) first.
   $3^2 = 9$
2. Do the **M**ultiplications next.
   $20 \times 3 = 60$
   $5 \times 9 = 45$
3. Do the **S**ubtraction last.
   $60 - 45 = 15$

Don't try to do more than one operation on each line of working.

## Worked example

This formula is used in physics to calculate distance.

$D = ut - 5t^2$

$u = 20$

$t = 3$

Work out the value of $D$. **(3 marks)**

$D = 20 \times 3 - 5 \times 3^2$

$\quad = 20 \times 3 - 5 \times 9$

$\quad = 60 - 45$

$\quad = 15$

---

## Now try this

**1** Here is a trapezium.

The area of the trapezium is given by the formula Area $= \frac{1}{2}(a + b)h$

Work out the area when $h = 4.5$, $a = 6.5$ and $b = 9.5$ **(3 marks)**

$a$ cm

$h$ cm

$b$ cm

**2** Here is a formula: $D = b^2 - 4ac$

Work out the value of $D$ when $a = 2$, $b = 5$ and $c = -3$ **(3 marks)**

Substitute all the values before you do any calculations.

# Writing formulae

You can write a rule given in words as a word formula or as a formula using algebra.

This label gives instructions for working out the cooking time of a chicken.

| FREE-RANGE CHICKEN | | |
|---|---|---|
| WEIGHT (KG) | PRICE PER KG | COOKING INSTRUCTIONS |
| 1.8 | £3.95 | Cook at 170°C for 25 minutes per kg plus half an hour |

You can write the cooking instructions as a word formula.

**Cooking time in minutes = 25 × weight in kg + 30**

You need to give units when you are describing the quantities in a formula.
If the cooking time is in hours then this formula would give you a very crispy chicken!

You can also write this formula using algebra.

$T = 25w + 30$, where $T$ is the cooking time in minutes and $w$ is the weight in kg.

When you write a formula using algebra you need to explain what each letter means.

## Worked example

Chloe buys $a$ pens costing 25 pence each and $b$ pencils costing 15 pence each. Write a formula for the total cost $T$ pence. **(2 marks)**

$T = 25a + 15b$

25$a$ means 25 × $a$. You should make sure your formula is simplified as much as possible. Don't write any units (like pence) in your formula.

## Worked example

The cost of hiring a car can be worked out using this formula.

| Cost = £80 + 50p per mile |
|---|

Write a formula for the cost £$C$ of hiring a car which is driven for $m$ miles. **(2 marks)**

$C = 80 + 0.5m$

You need to make sure all your values are in the same units. $C$ is the cost in pounds, so convert 50p into pounds. 50p = £0.50, so you need to multiply the number of miles by 0.5

## Worked example

The diagram shows a regular hexagon.

Write a formula for the perimeter of the hexagon $P$ in terms of $s$. **(1 mark)**

$P = 6s$

← s →

The perimeter is the distance all the way around a shape.
All the sides of a regular shape are the same length. There are 6 sides so:
Perimeter = 6 × length of one side
$P = 6s$

You can find out more about perimeter on page 80.

## Now try this

 *tick*

1   Rulers cost 50 pence each and pens cost 35 pence each. Which of these is a formula for the total cost, £$T$, when buying $x$ rulers and $y$ pens?
Circle your answer. **(1 mark)**

$T = 35x + 50y$     $T = 0.35x + 0.5y$

$T = 50x + 35y$     $T = 0.5x + 0.35y$

2   This shape is made from 3 identical squares.

Worked solution video

← s →

(a) Write a formula for the area, $A$, of the shape. **(1 mark)**

(b) Work out the area of the shape when $s = 5$ cm. **(1 mark)**

# Expanding brackets

Expanding brackets is sometimes called **multiplying out** brackets.

You need to be extra careful if there are negative signs outside the brackets.

$$-2 \times x = -2x$$
$$-2(x - y) = -2x + 2y$$
$$-2 \times -y = 2y$$

**Golden rule**

You have to multiply the expression outside the brackets by everything inside the brackets.

$$4n \times n = 4n^2$$
$$4n(n + 2) = 4n^2 + 8n$$
$$4n \times 2 = 8n$$

For a reminder about multiplying expressions look back at page 23.

Multiplying negative terms is like multiplying negative numbers.

When both terms are + or both terms are – the answer is **positive**.
$$-2a \times -a = +2a^2$$

When one term is + and one term is – the answer is **negative**.
$$-10p \times 5q = -50pq$$

---

Sometimes you have to **expand and simplify**. This means 'multiply out the brackets and then collect like terms'.

$$6 \times m = 6m \qquad 4 \times 3 = 12$$
$$6(m + 2) + 4(3 - m) = 6m + 12 + 12 - 4m$$
$$6 \times 2 = 12 \qquad 4 \times -m = -4m \qquad = 2m + 24$$
$$6m - 4m = 2m \qquad 12 + 12 = 24$$

Remember that any negative signs belong to the term on their right.

Look back at page 22 for a reminder about collecting like terms.

---

**Worked example**

(a) Expand $5(2y - 3)$     **(2 marks)**

$5(2y - 3) = 10y - 15$

(b) Expand and simplify $2(3x + 4) - 3(4x - 5)$     **(3 marks)**

$2(3x + 4) - 3(4x - 5)$
$= 6x + 8 - 12x + 15$
$= -6x + 23$

For part (b), be careful with the second bracket.
$$-3 \times 4x = -12x$$
$$-3 \times -5 = +15$$
The question says 'expand and simplify' so remember to collect any like terms after you have multiplied out the brackets.

---

**Now try this**

1 Expand (a) $3(y - 6)$   **(2 marks)**   (b) $m(m + 7)$   **(2 marks)**

2 Expand and simplify
   (a) $5(a - 2b) + 4(2a + b)$     **(3 marks)**
   (b) $2(4w + 3) - 3(3w - 5)$     **(3 marks)**
   (c) $2m(m + 9) - m(m - 4)$     **(3 marks)**

Be careful when there is a minus sign in front of the second bracket.
$$-3 \times 3w = -9w \text{ and}$$
$$-3 \times -5 = 15$$

# Factorising

**Factorising** is the opposite of expanding brackets.

$$2x(x + 3) = 2x^2 + 6x$$

You need to look for the largest factor you can take out of every term in the expression. This is what you need to write outside the brackets.

You have to factorise expressions as much as possible.

$10a^2 + 5a = 5(2a^2 + a)$

This expression has only been **partly factorised**.

$10a^2 + 5a = 5a(2a + 1)$

This expression has been **completely factorised**.
The two **factors** are $5a$ and $(2a + 1)$

## To factorise ...

1.  Look for the **largest factor** you can take out of every term.
    $12x - 8$
    The largest factor is 4.

2.  Write this factor outside the brackets. Write the sign (+ or −) from the expression inside the brackets.
    $12x - 8 = 4(\quad - \quad)$

3.  Work out what you need to multiply the factor by to get each term in the original expression.
    $12x - 8 = 4(3x - 2)$

4.  **Check** your answer by expanding the brackets. You should get back to the original expression.
    $4 \times 3x = 12x$
    $4(3x - 2) = 12x - 8 ✓$
    $4 \times -2 = -8$

## Problem solved!

When you write an expression with brackets both parts of the factorised expression are factors:

$6x^2 + 12xy = 6x(x + 2y) = \underline{3x}(2x + 4y)$
$\qquad\qquad\quad = \underline{x}(6x + 12y)$

So the three underlined expressions above are all factors of the original expression.

> You will need to use problem-solving skills throughout your exam – **be prepared!**

## Worked example

Circle all of the expressions that are factors of $6x^2 + 12xy$ **(2 marks)**

$$2x + y \qquad \boxed{x + 2y} \qquad 6y$$

$$y \qquad \boxed{x} \qquad \boxed{3x}$$

## Worked example

(a) Factorise $y^2 + 3y$ **(1 mark)**
$y^2 + 3y = y(y + 3)$

(b) Factorise fully $2p^2 - 4p$ **(2 marks)**
$2p^2 - 4p = 2p(p - 2)$

(a) **Check it!**
$y \times y = y^2 ✓ \quad y \times 3 = 3y ✓$

(b) The question says 'factorise fully'. This means you have to take the largest factor outside the brackets.
$2p^2 - 4p = 2(p^2 - 2p)$ is not fully factorised.

## Now try this

1  Factorise
   (a) $6n + 18$ **(1 mark)**
   (b) $3p - 3$ **(1 mark)**

2  Factorise fully
   (a) $4t^2 + 8t$ **(2 marks)**
   (b) $3x^2 - 12x$ **(2 marks)**

3  Circle all the factors of $2p^2 - 4pq$ **(2 marks)**

$$2 \quad 4p \quad 2p \quad p + 2q \quad p - 2q$$

Worked solution video

# Linear equations 1

An **equation** is like a pair of scales. The equals sign tells you that the scales are **balanced**. The letter represents an unknown weight. You can solve the equation to find the value of the letter.

These scales represent the equation $x + 7 = 10$

You need to subtract 7 from both sides.

You need $x$ on its own on one side.

The solution of this equation is $x = 3$

As long as you do the same thing to both sides, the scales stay balanced.

To **solve** an equation you need to get the letter on its own on one side.
It is important to write your working neatly when you are solving equations.

Every line of working should have an $=$ sign in it.

Start a new line for each step. Do one operation at a time.

$$5x + 3 = 18 \quad (-3)$$
$$5x = 15 \quad (\div 5)$$
$$x = 3$$

Write down the operation you are carrying out. Remember to do the same thing to both sides of the equation.

$5x$ means $5 \times x$. You have to divide by 5 to get $x$ on its own.

This is the solution to the equation. Your final line of working should look like this.

## Worked example

(a) Solve $m + 6 = 15$  **(1 mark)**

$$m + 6 = 15 \quad (-6)$$
$$m = 9$$

(b) Solve $8p = 36$  **(1 mark)**

$$8p = 36 \quad (\div 8)$$
$$p = 4.5$$

In part (b) you need to do **two operations** to get $a$ on its own. Do them one at a time, and make sure you show your working.

## Worked example

(a) Solve $\frac{x}{3} = -5$  **(1 mark)**

$$\frac{x}{3} = -5 \quad (\times 3)$$
$$x = -15$$

(b) Solve $2a - 7 = 11$  **(2 marks)**

$$2a - 7 = 11 \quad (+7)$$
$$2a = 18 \quad (\div 2)$$
$$a = 9$$

## Now try this

*Pick btwn 2 o? 3*

**1** Solve
   (a) $3y = 24$  **(1 mark)**
   (b) $m + 8 = 25$  **(1 mark)**
   (c) $31 - k = 12$  **(1 mark)**

**2** Solve
   (a) $4n - 7 = 13$  **(2 marks)**
   (b) $2w + 11 = 32$  **(2 marks)**

**3** (a) Solve $\frac{n}{4} = 9$  **(1 mark)**

   (b) Solve $\frac{q}{11} = 7$  **(1 mark)**

Your answer doesn't have to be a whole number.

Worked solution video

# Linear equations 2

## Equations with brackets

Always start by multiplying out the brackets then collecting like terms.

For a reminder about multiplying out brackets have a look at page 28.

$2 \times 3y = 6y$

$2(3y + 5)$

$2 \times 5 = 10$

$2(3y + 5) = 22$

$6y + 10 = 22$  $(-10)$

$6y = 12$  $(\div 6)$

$y = 2$

Do one operation at a time. Write down the operation you are using at each step.

## Equations with the letter on both sides

To solve an equation you have to get the letter on its own on one side of the equation.

Start by collecting like terms so that all the letters are together.

You can add or subtract multiples of $x$ on both sides of the equation.

$4x + 26 = 2 - 2x$  $(+2x)$

$6x + 26 = 2$  $(-26)$

$6x = -24$  $(\div 6)$

$x = -4$

Remember to do the same thing to both sides of the equation.

### Worked example

Solve $7p + 2 = 5(p - 4)$  **(3 marks)**

$7p + 2 = 5(p - 4)$  (expand brackets)

$7p + 2 = 5p - 20$  $(-5p)$

$2p + 2 = -20$  $(-2)$

$2p = -22$  $(\div 2)$

$p = -11$

Start by expanding the brackets on the right-hand side. Then subtract $5p$ from both sides to get all the $p$ terms together.

**Check it!**

Substitute $p = -11$ into both sides of the equation:

LHS: $7(-11) + 2 = -75$

RHS: $5(-11 - 4) = -75$ ✓

### Worked example

Solve $\dfrac{10 - y}{5} = 3$  **(3 marks)**

$\dfrac{10 - y}{5} = 3$  $(\times 5)$

$10 - y = 15$  $(+y)$

$10 = 15 + y$  $(-15)$

$-5 = y$

Start by multiplying both sides by 5 to remove the fraction. You want the $y$ term to be positive so add $y$ to both sides of the equation.

You can write your answer in the form $-5 = y$. As long as the letter is on its own on one side then the equation is solved.

### Now try this

1  (a) Solve $33 - 5x = 8$

**(2 marks)**

   (b) Solve $4 - 6x = 5 - 8x$

**(3 marks)**

Worked solution video

A positive $x$ term is easier to work with. Start by adding $5x$ to both sides.

2  (a) Solve $3t - 7 = 5t + 15$

**(3 marks)**

   (b) Solve $2(4h + 3) = 2$  **(3 marks)**

3  (a) Solve $3y + 19 = 5(y - 2)$

**(3 marks)**

   (b) Solve $\dfrac{16 - 5m}{3} = 7$  **(3 marks)**

Expand the brackets then group the $y$ terms on one side.

# Number machines

A number machine is sometimes called a **function machine**. It takes an **input** and applies **operations** to it to produce an **output**. You might need to write equations to solve problems involving number machines in your exams.

## Using algebra

If you don't know the input for a number machine you can write it as $x$. You can then use algebra to write the output as an expression.

### Golden rule

If the output and the input of a number machine are **both blank** you need to write an **equation** to find the missing values.

### Worked example

Here is a number machine:

Input  Output

The output is equal to the input.
Work out the input. **(3 marks)**

| INPUT | OUTPUT |
|-------|--------|
| $x$ | $5x - 2$ |

$$5x - 2 = x$$
$$5x = x + 2$$
$$4x = 2$$
$$x = \tfrac{1}{2}$$

### Problem solved!

Write the input as $x$, then use the information given in the question to form an equation. You can solve your equation to find the value of $x$.

You will need to use problem-solving skills throughout your exam – **be prepared!**

Follow these steps when the output and input are both blank.

- Write the input as $x$.
- Work out an expression for the output.
- Use the information in the question to write an equation.
- Solve your equation to work out the value of $x$.

**Check it!**
$\tfrac{1}{2} \times 5 - 2 = 2\tfrac{1}{2} - 2 = \tfrac{1}{2}$ ✓

### Now try this

Here is a number machine.

Input 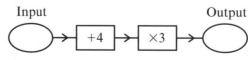 Output

The output is 4 times bigger than the input.
Work out the input.

**(3 marks)**

# Inequalities

You can use these symbols to describe **inequalities**. The fat end of the symbol always points towards the bigger number.

$>$ means 'is greater than'

$\geqslant$ means 'is greater than or equal to'

$<$ means 'is less than'

$\leqslant$ means 'is less than or equal to'

### Worked example

Write down all the integers which satisfy the inequality $3 < n \leqslant 6$.    **(1 mark)**

4, 5, 6

## Rounding error

You can use inequalities to show the range of possible values when a number has been **rounded**:

 The height of the London Eye is 140 m (to the nearest 10 m). You can write

135 m $\leqslant$ Height $<$ 145 m

The least possible value for the height is 135 m

You use 'less than' for the greatest possible value.

 $x = 4.7$ (1 d.p.) means

$4.65 \leqslant x < 4.75$

You use 'less than or equal to' for the least possible value.

---

## Inequalities on number lines

You can represent inequalities on a number line.

The open circle shows that $-1$ is **not** included.

The closed circle shows that 3 **is** included.

---

### Worked example

(a) Write down the inequality shown on the number line.

$x \leqslant 2$    **(1 mark)**

(b) Show the inequality $-1 < x \leqslant 4$ on the number line below.

   **(1 mark)**

(a) The circle is closed (filled in) so 2 **is** included.

(b) $-1 < x \leqslant 4$ is really two inequalities: $-1 < x$ and $x \leqslant 4$.

You need to show the values that satisfy **both** these inequalities at the same time.

You need to show all the values between $-1$ and 4. $-1$ is not included so draw an open circle at $-1$. 4 is included so draw a closed circle at 4.

---

### Now try this

1 Write down the inequalities shown on these number lines.

(a)

   **(1 mark)**

(b)

   **(1 mark)**

2 Show these inequalities on the number lines below.

(a) $-4 < x < 1$    **(1 mark)**    (b) $-2 \leqslant x < 5$    **(1 mark)**

# Solving inequalities

You can solve an inequality in exactly the same way as you solve an equation.

$2x - 3 \leq 15$     $(+ 3)$
$2x \leq 18$     $(\div 2)$
$x \leq 9$

## Solutions

The solution has the letter on its own on one side of the inequality and a number on the other side.

Solutions to inequalities
$x > 4$    $-2 < x$    $x \leq -\frac{3}{4}$

Not solutions to inequalities
$x \geq 20 + 3$    $2x < 10$    $x = 4$

## Watch out!

If you multiply or divide by a **negative number** when solving an inequality, you have to **reverse the direction** of the inequality sign. It's usually easier to eliminate negative terms by **adding**.

## Worked example

Solve the inequality
(a) $2x + 1 < 10$    $(- 1)$    **(2 marks)**
$2x < 9$     $(\div 2)$
$x < \frac{9}{2}$ or $x < 4\frac{1}{2}$

(b) $5x < 2x - 6$    $(- 2x)$    **(2 marks)**
$3x < -6$     $(\div 3)$
$x < -2$

(c) $12 - 4x \geq 10$    $(+ 4x)$    **(3 marks)**
$12 \geq 10 + 4x$    $(- 10)$
$2 \geq 4x$     $(\div 4)$
$\frac{2}{4} \geq x$ or $\frac{1}{2} \geq x$

These are **inequalities** and not equations. You don't need to use an = sign anywhere in your answer.
Write down the operation you are using at each step and remember you have to do the same thing to both sides of the inequality.
Your answer can involve negative numbers or fractions.

When you're solving an inequality you should avoid multiplying or dividing by a **negative** number. Add $4x$ to both sides of the equation to make the $x$ term positive.

## Worked example

Write down all the integers that satisfy the inequality:
$-15 \leq 5n < 10$    **(3 marks)**

$-15 \leq 5n$    $5n < 10$    $(\div 5)$
$-3 \leq n$     $n < 2$
$n = -3, -2, -1, 0, 1$

This is really two inequalities:
$-15 \leq 5n$   and   $5n < 10$
Solve them separately, then write down all the integers which satisfy **both** inequalities at the same time.

## Now try this

1 Solve the following inequalities.
   (a) $4x + 3 \geq 23$    **(2 marks)**
   (b) $18 - 5x > -2$    **(3 marks)**

2 Find all the integers which satisfy each of these inequalities.
   (a) $-8 \leq 4x < 12$    **(3 marks)**
   (b) $-1 < 3x \leq 10$    **(3 marks)**

Worked solution video

# Sequences 1

A **sequence** is a pattern of numbers or shapes that follow a rule.

2, 4, 6, 8, 10... is a sequence of even numbers.

1, 4, 9, 16, 25... is a sequence of square numbers.

Each number in a number sequence is called a **term**.

You can continue a sequence of numbers by finding the rule to get from one term to the next.

 **Worked example**

(a) The numbers in this sequence increase by the same amount each time.

11   !5.   19   23   27   3!.

+4   +4

Work out the two missing numbers. **(2 marks)**

Everything in red is part of the answer.

---

 **Problem solved!**

You are given the third term. To work out the term **before** this you need to work **backwards** through the rule. You can use function machines to help you.

→ +6 → ÷2 →

← −6 ← ×2 ←

**Check it!**
$(14 + 6) \div 2 = 20 \div 2 = 10$
$(10 + 6) \div 2 = 16 \div 2 = 8$ ✓

You will need to use problem-solving skills throughout your exam – **be prepared!**

 **Worked example**

Here is a sequence. The third term of this sequence is 8.

...   ...   8   ...

The rule for this sequence is

> Add 6 to the previous term and then divide by 2

Work out the first term of this sequence.
**(3 marks)**

3rd term = 8
2nd term = 8 × 2 − 6 = 10
1st term = 10 × 2 − 6 = 14

---

## Generating sequences

You can work out the terms of a sequence by substituting the term number into the $n$th term. Here are some examples:

| $n$th term | $9 - 2n$ | $n^2 + 10$ |
|---|---|---|
| 1st term | $9 - 2 \times 1 = 7$ | $1^2 + 10 = 11$ |
| 2nd term | $9 - 2 \times 2 = 5$ | $2^2 + 10 = 14$ |
| ↓ | ↓ | ↓ |
| 8th term | $9 - 2 \times 8 = -7$ | $8^2 + 10 = 74$ |

 **Worked example**

The $n$th term of a sequence is $50 - n^2$.
Work out the first term of the sequence that is negative. **(2 marks)**

7th term = $50 - 7^2 = 1$
8th term = $50 - 8^2 = -14$
The 8th term is the first negative term.

---

 **Now try this**

 **1** The first five terms of a sequence are

3   5   9   15   23

Write down the next two terms of this sequence.
**(1 mark)**

**2** The $n$th term of a sequence is $7n - 5$
(a) Work out the 8th term of this sequence.
**(1 mark)**
(b) Work out the first term of this sequence that is greater than 70. **(1 mark)**

**3** The rule for generating a sequence is 'add two consecutive terms to get the next term'.
Here are the first five terms.
2   3   5   8   13
Write down the next three terms of this sequence.  **(1 mark)**

This type of sequence is called a **Fibonacci-type** sequence.

# Sequences 2

An **arithmetic** or **linear sequence** is a sequence of numbers where the difference between consecutive terms is **constant**. In your exam, you might need to work out the $n$th term of a sequence.

## Worked example

Here is a sequence of numbers:

$1 \xrightarrow{+4} 5 \xrightarrow{+4} 9 \xrightarrow{+4} 13 \xrightarrow{+4} 17$

Work out a formula for the $n$th term of the sequence. **(2 marks)**

$4n$:      4    8    12    16    20    ...

$\boxed{-3}$         $\boxed{-3}$

Sequence:   1    5    9    13    17    ...

$n$th term $= 4n - 3$

You can use the formula for the $n$th term to generate any term in the sequence:
20th term $= 4 \times 20 - 3 = 80 - 3 = 77$

Start by finding the constant difference between consecutive terms in the sequence. For this sequence the common difference is 4.
Write out the sequence with general term $4n$, and compare this with the sequence given in the question.
To get each term in the sequence you have to **subtract 3** from each term in the $4n$ sequence. So the general term of the sequence is $4n - 3$.

**Check it!**
You can check your answer by substituting values of $n$ into your $n$th term.
1st term: when $n = 1$,
$4n - 3 = 4 \times 1 - 3 = 1$ ✓
2nd term: when $n = 2$,
$4n - 3 = 4 \times 2 - 3 = 5$ ✓

## Is 99 in this sequence?

The value of $n$ in the $n$th term has to be a **positive whole number.**
You can use this fact to check whether a number is a term in the sequence.

**Method 1:**
Try some different values of $n$:
$n = 25 \rightarrow 4n - 3 = 97$
$n = 26 \rightarrow 4n - 3 = 101$
You can't use a value of $n$ between 25 and 26 so 99 is **not** a term in the sequence.

**Method 2:**
Write an equation:
$4n - 3 = 99$
$4n = 102$
$n = 25.5$
25.5 is not a whole number so 99 is **not** a term in the sequence.

## Now try this

Worked solution video

**1** Here are the first five terms of an arithmetic sequence.
3    7    11    15    19
(a) Write down an expression for the $n$th term. **(2 marks)**
Karen says that 89 is a term in the sequence.
(b) Is she right? Give reasons for your answer. **(2 marks)**

Write down some terms in the sequence that are close to 89. Remember to write a conclusion to answer the question.

**2** Here is a sequence:
26    23    20    17     ...
(a) Which of these is a formula for the $n$th term of the sequence? Circle your answer. **(1 mark)**
$3n + 26$   $-3n + 26$   $3n + 29$   $-3n + 29$
(b) Work out the first term of this sequence that is negative. **(1 mark)**

Make sure you write the **term** and not the term number.

# Coordinates

You can use coordinates to describe the positions of points on a grid.

This point is to the left of O on the *x*-axis. So the *x*-coordinate is negative.

The vertical axis is labelled *y*.

The first number in a coordinate pair describes the horizontal position. The second number describes the vertical position.

The horizontal axis is labelled *x*.

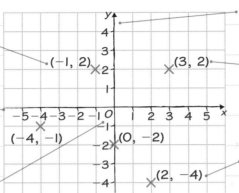

The point O is called the origin and has coordinates (0, 0).

You can use negative numbers to describe points below O on the *y*-axis.

---

## Worked example

(a) Write down the coordinates
   (i) of the point *A*          **(1 mark)**
   (2, 3)
   (ii) of the point *B*.        **(1 mark)**
   (2, −2)

(b) On the grid, plot the point *D* so that *ABCD* is a rectangle.          **(1 mark)**

Everything in red is part of the answer.

A rectangle has four right angles. Plot point *D* level with point *A* and vertically above point *C*.

---

## Midpoints

A **line segment** is a short section of a straight line.

The midpoint of a line segment is exactly halfway along the line. You can find the midpoint if you know the coordinates of the ends.

To find the midpoint, add the *x*-coordinates and divide by 2 and add the *y*-coordinates and divide by 2.

$$\text{Midpoint} = \left( \frac{x_1 + x_2}{2}, \frac{y_1 + y_2}{2} \right)$$

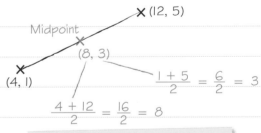

$$\frac{1 + 5}{2} = \frac{6}{2} = 3$$

$$\frac{4 + 12}{2} = \frac{16}{2} = 8$$

You need to be confident in finding midpoints if you're aiming for a top grade.

---

## Now try this

*ABCD* is a rectangle.
(a) Work out the coordinates of *D*.          **(1 mark)**
(b) Work out the length of *AB*.          **(1 mark)**
(c) *E* is the centre of the rectangle.
   Work out the coordinates of *E*.          **(2 marks)**

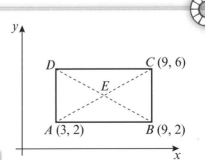

*E* is the midpoint of *AC* and of *BD*.

37

# Gradients of lines

The **gradient** of a straight-line graph measures how steep the line is. You can work out

the gradient by drawing a triangle and using this rule: Gradient = $\dfrac{\text{distance up}}{\text{distance across}}$

## Worked example

This scatter graph shows the relationship between the budgets of some films and the amounts of money they made at the box office in their opening weekends. A line of best fit has been drawn on the scatter graph.

Budget (£ million)

Everything in red is part of the answer.

(a) Work out the gradient of the line of best fit. **(1 mark)**

Gradient = $\dfrac{\text{distance up}}{\text{distance across}} = \dfrac{6}{60} = 0.1$

(b) Amir says 'Spending an extra £10 million on a film is likely to increase the amount of money taken during its opening weekend by £1 million.' Does the graph support this statement? **(1 mark)**

Yes. The gradient of the line of best fit is 0.1, so the amount of money taken during the opening weekend increases by approximately £0.1 million for each additional £1 million of budget.

To work out the gradient of a line you need to draw a triangle.

Write the distance across and the distance up.

**Watch out** for the scales on the axes:

Distance across = 80 – 20 = 60

Distance up = 10 – 4 = 6

There is more on scatter graphs on page 120.

### Top triangle tips!

1. Draw one side of your triangle on a large grid line as you are less likely to make a mistake in your calculations.

2. Use a large triangle as this means your calculations are more accurate.

3. Don't just count grid squares. Use the scale to work out the distance across and the distance up.

## Positive or negative?

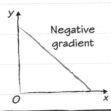

Gradient = 0

Negative gradient

If the gradient is negative then one value decreases as the other value increases.

## Now try this

Use the **scale** when calculating the gradient

The scatter graph shows a relationship between the amount of money a company spent on marketing each month and its revenue.

(a) Calculate the gradient of the line of best fit. **(1 mark)**

(b) Amy states that 'For every £10 000 spent on advertising, revenue increases by £20 000.'
Does the scatter graph support her statement? **(1 mark)**

Money spent (£1000s)

# Straight-line graphs 1

Here are two things you need to know about straight-line graphs.

 **1** If an equation is in the form $y = mx + c$, its graph will be a straight line.

$$y = -\frac{1}{2}x + 5$$

This number tells you the gradient of the graph.

The $y$-intercept of the graph is at $(0, 5)$.

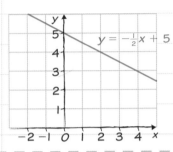

The gradient of the graph is $-\frac{1}{2}$.
This means that for every unit you go across, you go half a unit down.

**2** Use a table of values to draw a graph.

$y = 2x + 1$

| $x$ | $-1$ | $0$ | $1$ | $2$ |
|---|---|---|---|---|
| $y$ | $-1$ | $1$ | $3$ | $5$ |

→ $y = 2 \times 2 + 1 = 5$

Choose simple values of $x$ and substitute them into the equation to find the values of $y$.

Plot the points on your graph and join them with a straight line.

---

## Worked example

On the grid draw the graph of $x + y = 4$ for values of $x$ from $-2$ to $5$

**(3 marks)**

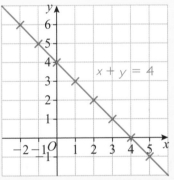

Everything in red is part of the answer.

| $x$ | $-2$ | $-1$ | $0$ | $1$ | $2$ | $3$ | $4$ | $5$ |
|---|---|---|---|---|---|---|---|---|
| $y$ | $6$ | $5$ | $4$ | $3$ | $2$ | $1$ | $0$ | $-1$ |

## Finding equations

If you have a graph you can find its equation by working out the gradient and looking at the $y$-intercept.

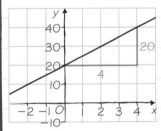

Draw a triangle to find the gradient.

$$\text{Gradient} = \frac{20}{4} = 5$$

The $y$-intercept is $(0, 20)$.

Put your values for gradient, $m$, and $y$-intercept, $c$, into the equation of a straight line, $y = mx + c$.

The equation is $y = 5x + 20$

---

You can rearrange the equation of this graph into the form $y = mx + c$ so it is a **straight line**.

$x + y = 4$ 　　　　 $(-x)$
　$y = -x + 4$ 　 $m = -1$ and $c = 4$

The gradient is $-1$ and the $y$-intercept is at $(0, 4)$. You could use this information to draw the graph, but it's safer to make a table of values. Make sure you plot **at least three** points, then join them with a straight line using a ruler.

## Now try this

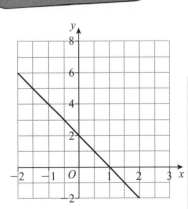

Find the equation of the straight line.

**(3 marks)**

Use $y = mx + c$. Draw a triangle to find the gradient, $m$. The graph slopes down so $m$ will be **negative**.

# Straight-line graphs 2

You can use **algebra** to find the equation of the line when you are given one point and the gradient, or two points.

## 1 Given one point and the gradient

| Substitute the gradient for $m$ in $y = mx + c$ | Gradient 2, passing through point (3, 7) |
|---|---|

↓

| Substitute the $x$- and $y$-values given into the equation | $y = 2x + c$<br>$7 = 2 \times 3 + c$<br>$7 = 6 + c$   $(-6)$<br>$c = 1$ |
|---|---|

↓

| Solve the equation to find $c$ | $y = 2x + 1$ |
|---|---|

↓

| Write out the equation | |
|---|---|

## 2 Given two points

| Draw a sketch showing the two points | Passing through points (8, 20) and (10, 30) |
|---|---|

↓

| Work out the gradient of the line with a triangle | Gradient $= \dfrac{10}{2} = 5$ |
|---|---|

↓

| Use method 1 (on the left) and one of the points given to find the equation | $y = 5x + c$<br>$30 = 5 \times 10 + c$<br>so $c = -20$<br><br>$y = 5x - 20$ |
|---|---|

---

## Parallel lines

**Parallel** lines have the same gradient.

These three lines all have a gradient of 1.

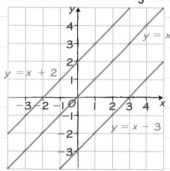

Parallel lines **never meet**.

### Problem solved!

Plan your strategy before you start:
1. Use the equation of line A to find the coordinates of points $P$ and $Q$.
2. Use the fact that line A is parallel to line B to write down the gradient of line B.
3. Use $y = mx + c$ to write down the equation of line B.

| You will need to use problem-solving skills throughout your exam – **be prepared!** |
|---|

### Worked example

Line A has equation $y = \frac{1}{2}x + 1$
Line B is parallel to line A.

Work out the equation of line B.    **(3 marks)**

$y = \frac{1}{2}x + 1$

$= \frac{1}{2} \times 6 + 1 = 3 + 1 = 4$

So Q has coordinates (6, 4)
So P has coordinates (0, 4)
Line B is parallel to line A so also has gradient $\frac{1}{2}$
So equation of line B is $y = \frac{1}{2}x + 4$

### Now try this

1  A straight line has gradient 6 and passes through the point (5, 10). Find the equation of the line.    **(2 marks)**

2  Find the equation of the straight line which passes through the points (3, 5) and (7, 13).    **(3 marks)**

| Find the gradient, $m$, then substitute $x = 3$ and $y = 5$ into $y = mx + c$. Solve the equation to find the value of $c$. |
|---|

# Real-life graphs

You can draw graphs to explain real-life situations. This graph shows the cost of buying some printed T-shirts from three different companies.

Be careful when reading scales on graphs. 10 small squares represent £50 so each small square represents £5.

| TERRY'S T-SHIRTS |
| --- |
| No minimum order |
| £5 per shirt |

| SHIRT-O-GRAPH |
| --- |
| Set-up cost £75 |
| £2.50 per shirt |

| PAM'S PRINTING |
| --- |
| Special offer 30 shirts for £100 |
| Additional shirts just £2 |

Terry's T-shirts would be cheapest if you were ordering 10 T-shirts.

## Worked example

This graph shows the cost, $C$ (£) of a taxi journey of length $m$ miles.

Everything in red is part of the answer.

(a) Use the graph to estimate the cost of a journey of 2.2 miles.     **(1 mark)**

£13

(b) The taxi company charges a fixed amount plus a cost per mile. Write down:
  (i)  the fixed amount
  (ii) the cost per mile.     **(2 marks)**

(i) £2     (ii) £5

> The additional cost per mile is the **gradient** of the graph.

## Interpreting the gradient

You might need to **interpret the gradient** of a straight-line graph:

✓ Straight-line graphs have a **constant gradient**.

✓ The gradient tells you how much the **vertical** variable increases when you increase the **horizontal** variable by one unit.

✓ Pay close attention to the scale on the axes – don't just count squares on the graphs.

There's more about gradients on page 38.

## Now try this

A mobile phone company charges a fixed monthly price plus a cost per minute.
(a) What is the fixed monthly charge?     **(1 mark)**
(b) What is the total cost if you use 400 minutes in one month?     **(1 mark)**
(c) The phone company claims that 'For every 10 minutes you only pay an extra 25p.'
    Is the phone company's claim true?
    Give a reason for your answer.     **(3 marks)**

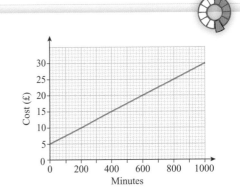

# Distance–time graphs

A **distance–time** graph shows how distance changes with time. This distance–time graph shows Jodi's run. The shape of the graph gives you information about the journey.

A horizontal line means no movement. Jodi rested here for 15 minutes.

The gradient of the graph gives Jodi's speed.

This is when Jodi turned around and started heading back home.

The horizontal scale might be marked in minutes or hours. Remember that there are 60 minutes in 1 hour.

Straight lines mean that Jodi was travelling at a constant speed.

At 13:15 Jodi was 1.4 miles from home.

Jodi sped up when she was nearly home. The graph is steeper here.

---

## Worked example

This distance–time graph shows Nick's bike ride.

(a) For how long did Nick rest in total?     **(1 mark)**

20 minutes

(b) During which section of his journey was Nick travelling fastest? Give a reason for your answer.
     **(2 marks)**

A. Because that is where the graph is steepest.

The flat sections of the graph are when Nick is resting. Read the scale on the graph carefully.

---

## Now try this

Worked solution video

Christina rode her bike to a friend's house. She rested once on the way, had coffee at her friend's house, then rode home.

(a) How long did Christina spend at her friend's house?     **(1 mark)**

(b) Work out Christina's average speed for her return journey.     **(2 marks)**

You can find the gradient of the graph, or you can use    Average speed = $\dfrac{\text{Distance}}{\text{Time}}$

There is more about speed on page 65.

# Rates of change

A distance–time graph shows the **rate of change** of distance with time. You can revise distance–time graphs on page 42. You can use graphs to show other rates of change. These garden ponds are being filled with water. The graphs show the rate of change of the **depth** of the water in each pond.

The narrower the section of pond, the quicker the water depth will increase.

---

## Worked example

Supraj recorded the rate of rainfall, in mm per hour, during one afternoon.

(a) At what time did it start to rain?

1330

(b) Supraj said, 'It stopped raining at 1400.'
Is Supraj correct? Give a reason for your answer.

Supraj is wrong. The rain was falling steadily from 1400 to 1600. It didn't stop raining until 1730.

 This graph shows the **rate** of rainfall. The rain was getting harder from 1330 to 1400. It then fell at a steady rate until 1600, when it started getting lighter.

---

## Velocity–time graphs

Velocity means **speed** in a certain direction. A velocity–time graph shows the rate of change of **velocity** with time. The **gradient** on a velocity–time graph tells you the **acceleration**. You need to recognise these four situations:

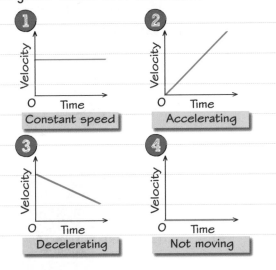

## Now try this

This is the velocity–time graph for a car journey.

Which of these statements are true and which are false? Give reasons for your answers.

**A** The car travelled at a constant speed between 0 and 5 seconds.

**B** The car travelled at a constant speed between 5 and 15 seconds.

**C** The car was accelerating between 0 and 5 seconds.

**D** The car was not moving between 5 and 15 seconds. **(4 marks)**

# Expanding double brackets

You need to be able to expand the **product** of two brackets. Use one of these three methods.

## 1 One-at-a-time

You can expand one bracket first, then the other:

Now expand and simplify as normal

$n(n + 3)$

$(n + 3)(n + 8) = n(n + 3) + 8(n + 3)$
$= n^2 + 3n + 8n + 24$
$= n^2 + 11n + 24$

$+ 8(n + 3)$

Remember to collect like terms

## 2 FOIL

The acronym **FOIL** tells you the four terms you need to find:

$2a^2 \quad -b^2$

$(2a + b)(a - b) = 2a^2 - 2ab + ab - b^2$
$= 2a^2 - ab - b^2$

$ab$

$-2ab$

First terms
Outer term
Inner terms
Last terms

Some people remember this as a 'smiley face'.

## 3 Grid method

You can draw a grid to help you expand the brackets:

$(x + 7)(x - 5) = x^2 - 5x + 7x - 35$
$= x^2 + 2x - 35$

Remember to collect like terms if possible.

|   | x | −5 |
|---|---|---|
| x | $x^2$ | $-5x$ |
| 7 | $7x$ | $-35$ |

The negative sign belongs to the 5. You need to write it in your grid.

### Worked example

Expand and simplify $(y + 2)(y + 9)$   **(2 marks)**

$(y + 2)(y + 9) = y^2 + 9y + 2y + 18$
$= y^2 + 11y + 18$

$y^2 \quad 18$

$(y + 2)(y + 9)$

$2y$

$9y$

This example shows the **FOIL** method. The question says expand and **simplify** so collect like terms at the end.

### Worked example

Expand and simplify $(3p - 4)^2$   **(2 marks)**

$(3p - 4)^2 = (3p - 4)(3p - 4)$
$= 9p^2 - 12p - 12p + 16$
$= 9p^2 - 24p + 16$

|   | 3p | −4 |
|---|---|---|
| 3p | $9p^2$ | $-12p$ |
| −4 | $-12p$ | 16 |

When you square a term in brackets you multiply the **whole bracket** by itself. Start by writing out:
$(3p - 4)^2 = (3p - 4)(3p - 4)$
This example shows the **grid method** but you could use any method to expand the brackets.
Remember that $3p \times 3p = 9p^2$ and be careful with the **negative signs**:
$-4 \times -4 = 16 \quad p \times -4 = -4p$

### Now try this

**1** Complete this working:
$(x + 6)(x + 3) = x^2 + 3x + \square x + \square$
$= x^2 + \square x + \square$   **(2 marks)**

**2** Expand and simplify
(a) $(x + 5)(x - 1)$ **(2 marks)**
(b) $(p - 6)^2$    **(2 marks)**

Worked solution video

# Quadratic graphs

**Quadratic** equations contain an $x^2$ term. Quadratic equations have **curved** graphs. You can draw the graph of a quadratic equation by completing a table of values.

> The **turning point** is the point where the direction of the curve changes.

## Worked example

(a) Complete the table of values for $y = 4x - x^2$. **(2 marks)**

| $x$ | −1 | 0 | 1 | 2 | 3 | 4 | 5 |
|-----|----|---|---|---|---|---|---|
| $y$ | −5 | 0 | 3 | 4 | 3 | 0 | −5 |

(b) On the grid, draw the graph of $y = 4x - x^2$. **(2 marks)**
(c) Write down the coordinates of the turning point. **(1 mark)**

(2, 4)

When $x = -1$: $4 \times (-1) - (-1)^2 = -4 - 1 = -5$
When $x = 4$: $4 \times 4 - 4^2 = 16 - 16 = 0$
Plot your points carefully on the graph and join them with a **smooth** curve.

**Check it!**
All the points on your graph should lie on the curve. If one of the points doesn't fit then double-check your calculation.

Everything in red is part of the answer.

## Drawing quadratic curves

- ✓ Use a sharp pencil.
- ✓ Plot the points carefully.
- ✓ Draw a smooth curve that passes through every point.
- ✓ Label your graph.
- ✓ Shape of graph will be either ∪ or ∩

## Drawing a smooth curve

It's easier to draw a smooth curve if you turn your graph paper so your hand is **inside** the curve.

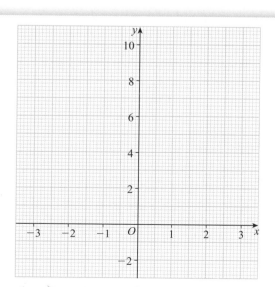

## Now try this

Worked solution video

(a) Complete this table of values for $y = x^2 + 2$ **(2 marks)**

| $x$ | −3 | −2 | −1 | 0 | 1 | 2 | 3 |
|-----|----|----|----|---|---|---|---|
| $y$ |    | 6  |    | 2 | 3 | 6 |   |

(b) Draw the graph of $y = x^2 + 2$ for $x = -3$ to $x = 3$ **(2 marks)**
(c) Use your graph to find the value of $y$ when $x = 2.5$ **(1 mark)**

# Using quadratic graphs

You might need to read values off a quadratic graph. You can do this by looking at the points where the graph crosses the $x$-axis, or by drawing a horizontal line on your graph.

## Worked example

(a) Complete the table of values for $y = x^2 - x - 3$

**(2 marks)**

| $x$ | $-2$ | $-1$ | $0$ | $1$ | $2$ | $3$ |
|---|---|---|---|---|---|---|
| $y$ | $3$ | $-1$ | $-3$ | $-3$ | $-1$ | $3$ |

(b) On the grid, draw the graph of $y = x^2 - x - 3$

**(2 marks)**

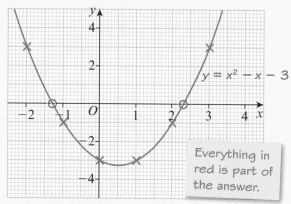

$y = x^2 - x - 3$

Everything in red is part of the answer.

> This is an example of a **quadratic graph**. Its equation contains an $x^2$ term. You should always use a table of values to draw quadratic graphs. Plot the points from the table and join them with a smooth curve. Then label your graph with its equation.

(c) Use your graph to write down the solutions of
$x^2 - x - 3 = 0$

**(2 marks)**

$x = -1.3$ and $x = 2.3$

> The solutions of the equation $x^2 - x - 3 = 0$ are the $x$-coordinates at the points where the graph crosses the $x$-axis. These are sometimes called the **roots** of the equation. You should always read graphs accurate to the nearest small square. So on this graph your answers will be accurate to 1 decimal place.

## Now try this

This is the graph of $y = x^2 - 4x - 1$
(a) Write down the values of $x$ where the graph crosses the $x$-axis.

**(2 marks)**

(b) Write down the values of $x$ when $y = 8$

**(2 marks)**

> Draw a horizontal line on the graph at $y = 8$.
> Read down to the $x$-axis at the points where the line crosses the curve.
> Read the graph correct to the nearest small square.
> These values of $x$ are the solutions to the quadratic equation $x^2 - 4x - 1 = 8$

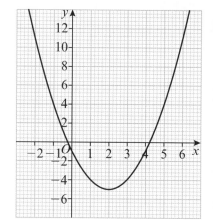

# Factorising quadratics

You need to be able to **factorise** quadratic expressions by writing them as the product of **two brackets**. Make sure you are really confident expanding **double brackets** before you start. You can revise this on page 44. You can also revise factorising simpler expressions on page 29.

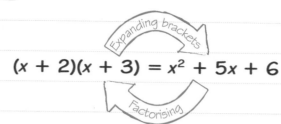

$$(x + 2)(x + 3) = x^2 + 5x + 6$$

*Expanding brackets*
*Factorising*

## Factorising $x^2 + bx + c$

You need to write the expression with **two brackets**. Look for two numbers whose **sum** is equal to $b$ and whose **product** is equal to $c$.

You need to find two numbers which add up to 7...     $5 + 2 = 7$

$$x^2 + 7x + 10 = (x + 5)(x + 2)$$

$5 \times 2 = 10$

... and multiply to make 10

You can use this table to help you find the two numbers.

| $b$ | $c$ | Factors |
|---|---|---|
| Positive | Positive | Both numbers positive |
| Positive | Negative | Bigger number positive and smaller number negative |
| Negative | Negative | Bigger number negative and smaller number positive |
| Negative | Positive | Both numbers negative |

## Worked example

Factorise $x^2 - x - 20$     **(2 marks)**

Factor pairs of 20:
1 and 20, 2 and 10, <u>4 and 5</u>
$x^2 - x - 20 = (x + 4)(x - 5)$
Check:
$(x + 4)(x - 5) = x^2 - 5x + 4x - 20$
$= x^2 - x - 20$ ✓

## Top tips

**1** Try factor pairs of the number part, $c$, when factorising $x^2 + bx + c$.

**2** Expand the brackets to check your answer.

## Difference of two squares

You can factorise expressions that are written as
(something)$^2$ − (something else)$^2$

Use this rule:
$a^2 - b^2 = (a + b)(a - b)$   **LEARN IT!**
$x^2 - 36 = x^2 - 6^2$
$= (x + 6)(x - 6)$

36 is a square number.
$36 = 6^2$ so $a = x$ and $b = 6$

## Worked example

Factorise $x^2 - 9$     **(2 marks)**

$x^2 - 9 = (x + 3)(x - 3)$
Check:
$(x + 3)(x - 3) = x^2 - 3x + 3x - 9$
$= x^2 - 9$ ✓

Keep an eye out for the difference of two squares. $9 = 3^2$ so $a = x$ and $b = 3$.

## Now try this

Worked solution video

1 Factorise
   (a) $x^2 + 6x + 8$     **(2 marks)**
   (b) $x^2 - 10x + 16$     **(2 marks)**

2 Factorise
   (a) $x^2 - 144$     **(2 marks)**
   (b) $x^2 - 49$     **(2 marks)**

# Quadratic equations

You need to be able to **solve** a quadratic equation **without a calculator** by factorising one side. If you need to solve a quadratic equation in your Foundation exam it will look something like this:

The left-hand side will be a **quadratic expression**

$$x^2 + 6x + 5 = 0$$

The right-hand side will be **zero**

Follow the steps below to solve the equation.

 **1** Factorise the left-hand side of the equation:

$(x + 5)(x + 1) = 0$

 **2** Set each factor equal to zero:

$(x + 5)(x + 1) = 0$

$(x + 5) = 0$　$(x + 1) = 0$

 **3** Solve the two linear equations to find the two solutions:

$x + 5 = 0$　$x + 1 = 0$

$x = -5$　　$x = -1$

For a reminder about factorising quadratic expressions look at page 47.

Write both solutions in the form $x = \boxed{\phantom{x}}$

---

## Worked example

Solve $x^2 + 8x - 9 = 0$　　**(3 marks)**

$(x + 9)(x - 1) = 0$

$x + 9 = 0$　　　$x - 1 = 0$

$x = -9$　　　　$x = 1$

1. Start by factorising the left-hand side of the equation. You can check by expanding the brackets:

$(x + 9)(x - 1) = x^2 - x + 9x - 9$
$= x^2 + 8x - 9$ ✓

2. Set each factor equal to zero.

3. Solve the linear equations to get **two** solutions.

**Check it!**
$(1)^2 + 8(1) - 9 = 1 + 8 - 9 = 0$ ✓
$(-9)^2 + 8(-9) - 9 = 81 - 72 - 9 = 0$ ✓

## Worked example

Solve $x^2 - 100 = 0$　　**(3 marks)**

$(x + 10)(x - 10) = 0$

$x + 10 = 0$　　　$x - 10 = 0$

$x = -10$　　　　$x = 10$

You can also use inverse operations:
$x^2 - 100 = 0$　　　　$(+ 100)$
$x^2 = 100$　　　　　$(\sqrt{\phantom{x}})$
$x = 10$ or $x = -10$

There are two answers because
$(-10)^2 = -10 \times -10 = 100$

### Two to watch

**1** When one solution is $x = 0$
$x^2 - 10x = 0$
$x(x - 10) = 0$
Solutions are $x = 0$ and $x = 10$

**2** Difference of two squares.
$x^2 - 25 = 0$
$(x - 5)(x + 5) = 0$
Solutions are $x = 5$ and $x = -5$

---

## Now try this

For parts (a) and (c) look at the blue box above.

1 (a) Factorise $x^2 + 5x + 6$　　**(2 marks)**
　(b) Write down the two solutions of
　　$x^2 + 5x + 6 = 0$　　**(1 mark)**

Both solutions are **negative**.

2 Solve
　(a) $x^2 - 5x = 0$　　**(2 marks)**
　(b) $x^2 + 3x - 28 = 0$　　**(3 marks)**
　(c) $x^2 - 144 = 0$　　**(3 marks)**

# Cubic and reciprocal graphs

You might need to **draw** or **interpret** cubic and reciprocal graphs in your exam. You can use a **table of values** to draw any graph, but it helps if you know what the general shape of the graph is going to be.

**① Cubic graphs**

Graphs that contain an $x^3$ term and no higher powers of $x$ are called cubic graphs. Here are two examples:

$y = x^3 + 2x^2$

$y = 4 - x^3$

**② The reciprocal graph**

The graph of $y = \frac{1}{x}$ is called the reciprocal graph.

$y = \frac{1}{x}$

The graph gets closer and closer to the *x*- and *y*-axes but never touches them.

---

## Worked example

(a) Complete the table of values for $y = x^3 - 4x - 3$

| $x$ | $-2$ | $-1$ | $0$ | $1$ | $2$ | $3$ |
|-----|------|------|-----|-----|-----|-----|
| $y$ | $-3$ | $0$ | $-3$ | $-6$ | $-3$ | $12$ |

**(2 marks)**

(b) On the grid, draw the graph of $y = x^3 - 4x - 3$
for $-2 \leqslant x \leqslant 3$ **(2 marks)**

(c) (i) Estimate the value of $x$ when $y = 6$. **(1 mark)**
2.7

(ii) Comment on the accuracy of your estimate.
**(1 mark)**

The estimate is not very accurate because it is based on reading off a graph.

$y = x^3 - 4x - 3$

$y = 6$

Everything in red is part of the answer.

This is a **cubic graph** with a **positive** coefficient of *x*. If you recognise the shape of the graph then it's easier to tell if you've plotted your coordinates correctly.

---

## Now try this

**A**

**B**

**C**

**D**

**E**

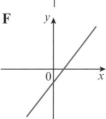

**F**

Write down the letter of the graph which could have the equation

(a) $y = -4x - 2$

(b) $y = x^3 - 3x + 4$

(c) $y = \frac{1}{x}$

(d) $y = 8x - 15 - x^2$

(e) $y = 2x - 3$

(f) $y = x^2 + 2x - 8$ **(6 marks)**

# Simultaneous equations

Simultaneous equations have two unknowns. You need to find the values for the two unknowns which make **both** equations true.

## Algebraic solution

1. Number each equation.

2. If necessary, multiply the equations so that the coefficients of one unknown are the same.

3. Add or subtract the equations to **eliminate** that unknown.

4. Once one unknown is found use substitution to find the other.

5. Check the answer by substituting both values into the original equations.

$$3x + y = 20 \quad ①$$
$$x + 4y = 14 \quad ②$$
$$12x + 4y = 80 \quad ① \times 4$$
$$- (x + 4y = 14) \quad -②$$
$$\overline{\phantom{xxxx} 11x = 66}$$
$$x = 6$$
Substitute $x = 6$ into ①:
$$3(6) + y = 20$$
$$18 + y = 20$$
$$y = 2$$
Solution is $x = 6$, $y = 2$.
Check: $x + 4y = 6 + 4(2) = 14$ ✓

## Worked example

Solve the simultaneous equations
$$6x + 2y = -3 \quad ①$$
$$4x - 3y = 11 \quad ②$$ **(4 marks)**

$$18x + 6y = -9 \quad ① \times 3$$
$$+ \ 8x - 6y = 22 \quad ② \times 2$$
$$\overline{\phantom{xxx} 26x = 13}$$
$$x = \tfrac{1}{2}$$
Substitute $x = \tfrac{1}{2}$ into ①:
$$6\left(\tfrac{1}{2}\right) + 2y = -3$$
$$3 + 2y = -3$$
$$2y = -6$$
$$y = -3$$

## Easier eliminations

You can save time by choosing the right unknown to eliminate. Look for one of these:

**1** If an unknown appears **on its own** in one equation you only need to multiply one equation to eliminate that unknown.

**2** If an unknown has **different signs** in the two equations you can eliminate by **adding**.

Multiply both equations by a whole number to make the coefficients of $y$ the same.

**Check it!**
Always use the equation you **didn't** substitute into to check your answer:
$$4x - 3y = 4\left(\tfrac{1}{2}\right) - 3(-3) = 2 + 9 = 11 ✓$$

## Graphical solution

You can solve these simultaneous equations by drawing a graph.
$$x - y = 1 \qquad x + 2y = 4$$
The coordinates of the point of intersection give the solution to the simultaneous equations.
The solution is $x = 2$, $y = 1$

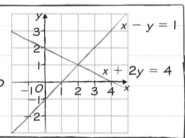

## Now try this

1 Solve the simultaneous equations
$$3x - 2y = 12$$
$$x + 4y = 11$$ **(3 marks)**

Draw a coordinate grid from −3 to 5 in both directions.

2 By drawing two suitable straight lines on a coordinate grid, solve these simultaneous equations.
$$3y + 2x = 6$$
$$y = 2x - 2$$ **(4 marks)**

Worked solution video

# Rearranging formulae

Most formulae have one letter on its own on one side of the formula. This letter is called the **subject** of the formula.

$P = 2l + 2w$    $P$ is the subject

$A = \frac{1}{2}bh$      $A$ is the subject

You can use algebra to change the subject of a formula. This is like solving an equation.

If you need to find a missing value which is not the subject in a formula:

1. Substitute any values you know into the formula.

For a reminder about substitution have a look at page 25.

2. Solve the equation to find the missing value.

## Worked example

Alicia works $h$ hours of normal time and $v$ hours of overtime each week. She is paid £$P$. She uses this formula to work out her pay:
$P = 8h + 12v$
Last week Alicia worked 32 hours of normal time and was paid £328. How many hours of overtime did she work?       **(3 marks)**

$328 = 8 \times 32 + 12v$
$328 = 256 + 12v$      $(-256)$
$72 = 12v$            $(\div 12)$
$6 = v$

Alicia worked 6 hours of overtime.

To make $p$ the subject of this formula you have to do the same thing to **both sides** of the formula until you have $p$ on its own on one side.

$N = 2p + 3q^2$  $(-3q^2)$ ← Subtract any terms you don't need.

 You have to divide **everything** on this side by 2.

$N - 3q^2 = 2p$      $(\div 2)$ ← $2p$ means $2 \times p$ so divide both sides by 2 to get $p$ on its own.

$\dfrac{N - 3q^2}{2} = p$

## Worked example

Rearrange this formula to make $Q$ the subject.
$R = 1 - 5Q$              **(2 marks)**

$R = 1 - 5Q$      $(+5Q)$
$R + 5Q = 1$        $(-R)$
$5Q = 1 - R$        $(\div 5)$
$Q = \dfrac{1 - R}{5}$

You want the term containing $Q$ to be positive. Start by adding $5Q$ to both sides.

## Worked example

Rearrange this formula to make $h$ the subject.
$A = \frac{1}{2}bh$              **(2 marks)**

$A = \frac{1}{2}bh$      $(\times 2)$
$2A = bh$        $(\div b)$
$\dfrac{2A}{b} = h$

You need the new subject on its own on one side of the formula. It doesn't matter which side.

## Now try this

Worked solution video

1 (a) Rearrange this formula to make $Q$ the subject.
   $W = 7Q - 100$              **(2 marks)**

(b) Rearrange this formula to make $v$ the subject.
   $d = \dfrac{m}{v}$              **(2 marks)**

2 Rearrange this formula to make $n$ the subject.
   $p = 2(m - n)$              **(2 marks)**

Start by expanding the brackets.

For a reminder about expanding brackets look at page 28.

51

# Using algebra

You can sometimes use the information given in a question to write an equation. You might need to choose a letter to represent an unknown quantity you are trying to find.

## Worked example

Here are three number cards.

   A B C

Each card has a whole number written on the back.
The number on card **B** is twice the number on card **A**.
The number on card **C** is five more than the number on card **A**.
The sum of the numbers on all three cards is 37.
Work out the number on each card. **(4 marks)**

$B = 2A$  $C = A + 5$
$A + B + C = 37$
$A + 2A + (A + 5) = 37$
$4A + 5 = 37$  $(- 5)$
$4A = 32$  $(\div 4)$
$A = 8$
$B = 2 \times 8 = 16$  $C = 8 + 5 = 13$

## Problem solved!

You could try guessing the numbers on the cards, but it might take a long time. You can solve this problem quickly by writing your own equation.
- Use **A** to represent the number on card **A**.
- Write expressions for the numbers on cards **B** and **C** in terms of **A**.
- Add together these three expressions and set the total equal to 37. Solve this equation to find **A**.

Once you have calculated **A**, remember to work out the values of **B** and **C** as well.

**Check it!**
$8 + 16 + 13 = 37$ ✓

You will need to use problem-solving skills throughout your exam – **be prepared!**

---

## True or false?

You might need to **substitute** values into an algebraic statement to show that it is **false**. You only need to find one **counter-example** to show that a statement is false.

Try some different values of *n* and write down all your working. You should know the prime numbers up to 20.

## Worked example

*n* is a whole number.
Supraj says that $2n^2 + 3$ is always a prime number.
Give an example to show that he is wrong. **(2 marks)**

$n = 1$:  $2n^2 + 3 = 2 \times 1 + 3 = 5$  Prime
$n = 2$:  $2n^2 + 3 = 2 \times 4 + 3 = 11$  Prime
$n = 3$:  $2n^2 + 3 = 2 \times 9 + 3 = 21 = 3 \times 7$
  Not prime

---

Worked solution video

## Now try this

**1** These two rectangles have the same perimeter.

$x - 3$
A | $2x$
$x$ | B
$x + 4$

In the diagram, all of the measurements are in metres. Work out the width and height of rectangle **A**.
Show all of your working. **(4 marks)**

**2** Jill says that the expression $n^2 + 4n + 3$, where *n* is an integer, is never a multiple of 7.
Is Jill correct? Write down calculations to support your answer. **(2 marks)**

Start by writing expressions for the perimeter of each rectangle. The perimeter of rectangle A is:
$(x - 3) + 2x + (x - 3) + 2x$
Remember to answer the question at the end.

# Identities

An **identity** is something that is **always** true. The right-hand side will always be exactly equal to the left-hand side, no matter what values you substitute. You use the symbol '≡' to represent an identity. You might need to **show** that an identity is true in your exam.

> ### Golden rule
> An identity is not like an equation. Do not solve it using the balance method.
> Manipulate each side separately. ✓
> Apply the same operation to both sides. ✗

### Worked example

Show that $(n - 1)^2 + (n + 1)^2 \equiv 2(n^2 + 1)$

**(2 marks)**

$(n - 1)^2 + (n + 1)^2$
$\equiv (n - 1)(n - 1) + (n + 1)(n + 1)$
$\equiv n^2 - 2n + 1 + n^2 + 2n + 1$
$\equiv 2n^2 + 2$
$\equiv 2(n^2 + 1)$

If you need to **show** that an identity is true in an exam you should show every line of your working.

Start with the expression on the left-hand side. Use multiplying out, simplifying and factorising to work towards the expression on the right-hand side.

## Odd or even?

You can sometimes work out whether an algebraic expression will have an **odd** or an **even** value. If you multiply a whole number by an even number the answer is **always even**.

If $n$ is a whole number then:
- $2n$ is **always even**
- $3n$ could be **odd** or **even**
- $4n$ is **always even**
- ... and so on

### Worked example

$x$ is a whole number.
Circle the words that describe $7x$.    **(1 mark)**

always odd    always even    (could be odd or even)

### Problem solved!

If a question involves algebra you might be able to try some simple numbers to see what is going on:
$x = 1 \rightarrow 7x = 7 \rightarrow$ **odd**
$x = 2 \rightarrow 7x = 14 \rightarrow$ **even**

> You will need to use problem-solving skills throughout your exam – **be prepared!**

### Now try this

**1** Show that $(a + b)(a - b) \equiv a^2 - b^2$
   **(2 marks)**

**2** Circle the expression that is equivalent to $n^2 - 4$.    **(1 mark)**

   $(n - 2)^2$     $(n + 2)(n - 2)$

   $(n^2 - 2)^2$     $(n - 4)(n - 1)$

> There is more about expanding brackets and factorising on pages 44 and 47.

**3** $a$ is an odd number and $b$ is an even number. Tick a box to show whether each statement is true or false.

**(3 marks)**

| | True | False |
|---|---|---|
| $2a$ is always an even number | ☐ | ☐ |
| $3b$ is always an odd number | ☐ | ☐ |
| $a + b$ could be an even or an odd number | ☐ | ☐ |
| Half of $b$ could be an odd number | ☐ | ☐ |

# Problem-solving practice 1

About half of the questions in your Foundation GCSE exam will require you to **problem-solve, reason, interpret** or **communicate** mathematically. If you come across a tricky or unfamiliar question in your exam you can try some of these strategies:

- ☑ Sketch a diagram to see what is going on.
- ☑ Try the problem with smaller or easier numbers.
- ☑ Plan your strategy before you start.
- ☑ Write down any formulae you might be able to use.
- ☑ Use *x* or *n* to represent an unknown value.

---

**1** In the diagram, ABCD is a rectangle.

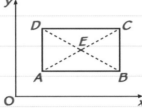

The length AB is twice the length AD.
E is the midpoint of the diagonals of the rectangle.
The coordinates of E are (30, 20).
Work out one set of possible coordinates for points A, B, C and D. **(4 marks)**

## Coordinates page 37

There are lots of different correct answers to this question. Start by picking coordinates for A. In order for the width of the rectangle to be twice its height, the horizontal distance from A to E must be twice the vertical distance from A to E.

**TOP TIP**

Choose simple numbers to make your calculations easier. The coordinates of the midpoint are multiples of 10, so try using multiples of 10 for the coordinates of the vertices.

---

**2** Dexter uses this formula to work out how long to cook a chicken.

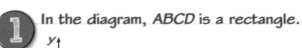

He works out that his chicken will take exactly 1 hour 33 minutes to cook.
Work out the weight of Dexter's chicken.

**(3 marks)**

## Rearranging formulae page 51

The units in the formula are minutes, so write 1 hour 33 minutes as a time in minutes. To find the weight of Dexter's chicken, substitute any values you know into the formula, and then solve the equation to find the missing value.

**TOP TIP**

You can represent the chicken's weight in kg as a single letter, e.g. *w*, in order to simplify your equation.

# Problem-solving practice 2

**3**

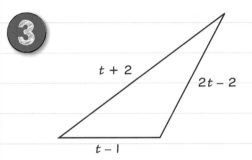

The perimeter of this triangle is 13 cm.

All lengths on the diagram are in cm.

Work out the value of $t$.    **(4 marks)**

### Linear equations 2 page 31
### Using algebra page 52

Use the information in the question to write an equation. Solve your equation to work out the value of $t$.

**TOP TIP**

Remember to simplify your equation by collecting like terms before solving.

---

**4** Here are the first four terms of an arithmetic sequence.

9, 15, 21, 27, ...

Is 65 a term in this sequence? Show working to support your answer.    **(3 marks)**

### Sequences 2 page 36

Work out the $n$th term of the sequence. Then work out consecutive terms of the sequence on either side of 65.

**TOP TIP**

You can use any successful strategy to answer a question, as long as you show your method. You could set the $n$th term equal to 65 and solve an equation to show that $n$ is not an integer.

---

**5** A train travels a distance of $y$ km in 3 minutes. Show that the average speed of the train is $20y$ km/h.    **(3 marks)**

### Formulae page 26    Speed page 65

**Watch out!** The first step of your working is to convert minutes into hours. Divide by 60 then simplify your fraction.

**TOP TIP**

Try changing $y$ into a number and calculating the average speed. It will give you an idea of how the formula works.

---

**6** (a) A straight line has a gradient of 8 and passes through the point (5, 10).

Find an equation of the line. Determine whether the point (10, 50) lies on the line.    **(3 marks)**

(b) Is your answer to part (a) reliable? Explain your answer.    **(1 mark)**

### Straight-line graphs 2 page 40

You could tackle this question using algebra, or by drawing a diagram to scale. Whichever method you use, make sure you write a conclusion.

**TOP TIP**

If you have to comment on **reliability** then think about the method you used. If you used a diagram your answer might be less reliable than if you used algebra.

# Percentages

'**Per cent**' means '**out of 100**'. You can write a percentage as a fraction over 100.

$$20\% = \frac{20}{100} = \frac{1}{5} \qquad 50\% = \frac{50}{100} = \frac{1}{2} \qquad 75\% = \frac{75}{100} = \frac{3}{4} \qquad 100\% = \frac{100}{100} = 1$$

## Percentages with a calculator

To find a percentage of an amount:

> Divide the percentage by 100
>
> ⬇
>
> Multiply by the amount

For example, 12% of 80 cm is 9.6 cm:

$12 \div 100 = 0.12$
$0.12 \times 80 = 9.6$

> First work out 15% of £120.
> Then subtract this from £120.

> You can also find 15% of £120 by working out 1% and then multiplying by 15.

### Worked example

A car rental company reduces its prices by 15% in a sale.
A car normally costs £120 per week to rent.
Work out the weekly rental cost of a car in the sale. **(3 marks)**

$15 \div 100 = 0.15$
$0.15 \times 120 = 18$
$120 - 18 = 102$

The car costs £102 per week in the sale.

---

To write one quantity as a percentage of another:

> Divide the first quantity by the second quantity
>
> ⬇
>
> Multiply your answer by 100

For example, 3 out of 12 yoghurts in a pack are strawberry.

$3 \div 12 = 0.25$
$0.25 \times 100 = 25$

So 25% of the yoghurts are strawberry.

### Worked example

In a year group of 96 students, 60 own a bicycle.
Express 60 as a percentage of 96. **(2 marks)**

$60 \div 96 = 0.625$
$0.625 \times 100 = 62.5$

62.5% of the students own a bicycle.

---

### Now try this

> 1% of £18 200 is £182, so 3% is 3 lots of £182.

 **1** Sam earns £18 200 a year.
He is given a pay rise of 3%.
How much is his pay rise? **(2 marks)**

 **2** A family recycle 8 kg of waste in a week.
2.8 kg of this waste is paper.
What percentage of the recycled waste is paper? **(2 marks)**

# Fractions, decimals and percentages

Here are three important facts about fractions, decimals and percentages:

 **1** You can convert a decimal to a percentage by multiplying by 100.

0.6 = 60%
0.03 = 3%

**2** You can write any percentage as a fraction with denominator 100.

$60\% = \frac{60}{100} = \frac{6}{10} = \frac{3}{5}$

Simplify your fraction as much as possible.

 **3** Remember these common fraction, decimal and percentage equivalents.

| Fraction | $\frac{1}{100}$ | $\frac{1}{10}$ | $\frac{1}{5}$ | $\frac{1}{4}$ | $\frac{1}{2}$ | $\frac{3}{4}$ |
|---|---|---|---|---|---|---|
| Decimal | 0.01 | 0.1 | 0.2 | 0.25 | 0.5 | 0.75 |
| Percentage | 1% | 10% | 20% | 25% | 50% | 75% |

You can arrange a list of fractions, decimals and percentages in order of size by changing them to the same type.

## Worked example

Carla has a bag of jelly beans.
15% of the jelly beans are strawberry.
$\frac{1}{4}$ of the jelly beans are pineapple.
$\frac{2}{5}$ of the jelly beans are apple.
The remaining jelly beans are cinnamon.
What percentage of the jelly beans are cinnamon? **(3 marks)**

$\frac{1}{4} = 25\%$
$\frac{2}{5} = 40\%$
15 + 25 + 40 = 80
100 − 80 = 20
20% of the jelly beans are cinnamon.

## Problem solved!

You might need to combine fractions, decimals and percentages in a word problem like this. Read the **whole question** before doing any working. The answer needs to be a **percentage** so the quickest way of answering this question would be to convert both fractions into percentages.

Remember that all of the jelly beans is 100%, so work out the sum of the other proportions then subtract that from 100% to find the percentage of jelly beans that are cinnamon.

 You will need to use problem-solving skills throughout your exam – **be prepared!**

## Now try this

**1** Write these percentages as fractions in their simplest form.
(a) 15% **(1 mark)**  $15\% = \frac{15}{100} = \frac{......}{20}$
(b) 68% **(1 mark)**

**2** Write these numbers in order, starting with the smallest.
0.42   $\frac{2}{5}$   36%   **(2 marks)**

Write all three numbers as decimals then compare.

**3** In a game, 5 players each put 12 counters on a table.
Sasha wins 35% of the counters.
Peter wins $\frac{2}{5}$ of the counters.
Haydon wins the remaining counters.
How many counters did Haydon win?
Show all of your working. **(3 marks)**

Worked solution video

# Percentage change 1

Here are two methods that you can use to increase or decrease an amount by a percentage.

## Method 1

Work out 26% of £280:

$\frac{26}{100} \times £280 = £72.80$

Subtract the decrease:

£280 − £72.80 = £207.20

**£280**

26% OFF

## Method 2

Use a multiplier:

100% + 30% = 130%

$\frac{130}{100} = 1.3$

So the multiplier for a 30% increase is 1.3:

400 g × 1.3 = 520 g

400 g
PLUS
**30%**
EXTRA

## Worked example

A football club increases the prices of its season tickets by 5.2% each year.

In 2011 a top-price season ticket cost £650.

Calculate the price of this season ticket in 2012. **(2 marks)**

$\frac{5.2}{100} \times £650 = £33.80$

£650 + £33.80 = £683.80

When working with money, you must give your answer to 2 decimal places.

**Check it!**

Choose an easy percentage which is close to 5.2%.

10% of £650 is £65, so 5% is £32.50.

£650 + £32.50 = £682.50, which is close to £683.80 ✓

## Calculating a percentage increase or decrease

Work out the amount of the increase or decrease

↓

Write this as a percentage of the original amount

Was £60
Now £39

60 − 39 = 21

$\frac{21}{60} = \frac{21}{60} \times 100\% = 35\%$

This is a 35% decrease.

A question might ask you to calculate a percentage **profit** or **loss** rather than an increase or decrease.

For a reminder about writing one quantity as a percentage of another, have a look at page 56.

## Now try this

Worked solution video

Reduction means decrease. Work out the decrease as a percentage of the original price.

 **1** In 2010, a business used 1755 reams of paper. In 2011 the same business reduced its paper usage by 23%. Work out how many reams of paper the business used in 2011. Give your answer to the nearest whole number. **(2 marks)**

 **2** A TV originally cost £520.
In a sale it was priced at £340.
What was the percentage reduction in the price? Give your answer to 1 decimal place. **(3 marks)**

# Percentage change 2

You need to be able to calculate a percentage increase or decrease **without a calculator.**

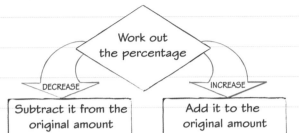

Work out the percentage

DECREASE → Subtract it from the original amount

INCREASE → Add it to the original amount

## Calculating percentages

You can use multiples of 1% and 10% to calculate percentages without a calculator.

Work out 12.5% of £600

10% of £600 is £60       600 ÷ 10 = 60

1% of £600 is £6         600 ÷ 100 = 6

0.5% of £600 is £3       6 ÷ 2 = 3

So, 12.5% of £600 is
£60 + £6 + £6 + £3 = £75

## Worked example

Which television is cheaper?

A £440  +20% VAT

B £550  SALE 2.5% OFF

Give reasons for your answer.    **(5 marks)**

Television A    10% = £44
                20% = £88
                £440 + £88 = £528

Television B

1% = £5.50
0.5% = £2.75
£550 − £5.50 − £5.50 − £2.75
= £536.25

Television A is cheaper.

## Problem solved!

The best way to give a reason for your answer is to show **all your working**. Make sure you lay out your working neatly and use the correct units.

**Television A**
1. Divide by 10 to calculate 10% of £440.
2. Multiply by 2 to work out 20%.
3. Add this to the original price.

**Television B**
1. Divide £550 by 100 to calculate 1%.
2. Divide by 2 to work out 0.5%.
3. Subtract two lots of £5.50 and one lot of £2.75 from the original price.

**Compare the prices.** It's always a good idea to write your answer as a sentence.

You will need to use problem-solving skills throughout your exam – **be prepared!**

---

Percentage increase and decrease questions come in lots of different forms.

### PERCENTAGE INCREASE

**INTEREST**
FIXED RATE
ISA 2.8%

**PAY RISE**

### PERCENTAGE DECREASE

**SALES**

10% off

**DEPRECIATION**

VALUE / TIME

---

## Now try this

Simon sees the same model of digital camera for sale in two different shops. Which shop is selling the camera at the cheaper price?
Give reasons for your answer.    **(5 marks)**

**CRUKS CAMERAS**
30% OFF NORMAL PRICE OF £245

**Spivs Cameras**
35% OFF NORMAL PRICE OF £270

# Ratio 1

**Ratios** are used to compare quantities.

The ratio of apples to oranges is 3 : 2
The ratio 3 : 2 is in its simplest form.

You can write a ratio as a fraction.
$\frac{2}{5}$ of the pieces of fruit are oranges.

3 + 2 = 5

The denominator is the sum of the parts in the ratio.

## Equivalent ratios

You can find equivalent ratios by multiplying or dividing by the same number.

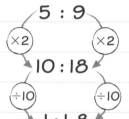

5 : 9

×2        ×2

10 : 18

÷10       ÷10

1 : 1.8

This equivalent ratio is in the form 1 : *n*
This is useful for calculations.

---

### Worked example

Jess and Simon are buying a new computer.
They pay in the ratio 2 : 3.
Jess pays £194.
How much does Simon pay?     **(3 marks)**

194 ÷ 2 = 97
3 × 97 = 291

Simon pays £291.

The order the people are written in is the same as the order of the numbers in the ratio. So 2 parts of the ratio represents the amount Jess paid. Divide £194 by 2 to work out how much each part of the ratio is worth, then multiply this value by 3 to find out how much Simon pays.

You can use a calculator, but remember to write down all your working.

### Worked example

A school has 220 students. 140 are female.
Work out the ratio of male to female students.
Give your answer in its simplest form.
**(2 marks)**

220 – 140 = 80

male : female

80 : 140

÷20        ÷20

4 : 7

## Simplest form

To write a ratio in its simplest form, find an equivalent ratio with the smallest possible whole number values.

Simplest form
5 : 1     10 : 9
2 : 3 : 4

Not simplest form
1 : 1.5 : 2   10 : 2
1 : 0.9

---

Worked solution video

### Now try this

1  (a) A bag contains white and pink marshmallows in the ratio 3 : 4. What fraction of the marshmallows are pink?  **(1 mark)**

   (b) In a second bag, $\frac{2}{5}$ of the marshmallows are white and the rest are pink. What is the ratio of white to pink marshmallows in the second bag?     **(1 mark)**

2  Write the ratio 84 : 120 in its simplest form.
     **(2 marks)**

3  A bag contains red counters and green counters. The ratio of red to green counters is 4 : 3. There are 32 red counters in the bag. What is the total number of counters in the bag?     **(3 marks)**

# Ratio 2

Ratio is used in lots of problem-solving questions. You need to be able to answer ratio questions like the ones on this page **without using a calculator.** Remember that you won't have one for Paper 1.

**Golden rule**

You can answer lots of ratio questions by working out what **one part** of the ratio represents.

## Worked example

Alexis, Nisha and Paul share a flat. One month their phone bill is £120.
They decide to split the bill in the ratio 3 : 5 : 2
How much does each person pay? **(3 marks)**

3 + 5 + 2 = 10
120 ÷ 10 = 12
3 × 12 = 36. Alexis pays £36.
5 × 12 = 60. Nisha pays £60.
2 × 12 = 24. Paul pays £24.

To divide a quantity in a given ratio:
1. Work out the total number of parts in the ratio.
2. Divide the quantity by this total.
3. Multiply your answer by each part of the ratio.

The order the people are written in is the same as the order of the numbers in the ratio. Alexis is first in the list, so 3 parts of the ratio represents the amount she pays.

**Check it!**
£36 + £60 + £24 = £120 ✓

## Worked example

Jamie and Chaaya took part in a sponsored swim to raise money for charity.
The ratio of Jamie's total to Chaaya's total is 5 : 7
Chaaya raised £12 more than Jamie.
How much money did they raise in total?
**(3 marks)**

7 − 5 = 2
2 parts = £12 so 1 part = £6
Jamie = £30
Chaaya = £42
Total = £30 + £42 = £72

## Problem solved!

Start by working out what one part of the ratio represents. Chaaya raised £12 more than Jamie so **two parts** of the ratio represent £12.
Therefore, **one part** of the ratio represents £6.
So Jamie raised 5 × £6 = £30
Chaaya raised 7 × £6 = £42
**Check it!**

30 : 42
÷6　　÷6
5 : 7

£42 − £30 = £12 ✓

You will need to use problem-solving skills throughout your exam – **be prepared!**

## Now try this

**1** Ruth, Sue and Tess share £400 between them.
Ruth receives £80 more than Sue.
The ratio of Ruth's share to Sue's share is 9 : 5
Work out how much Tess receives.
**(3 marks)**

**2** Terri mixed 300 g of rice with 240 g of fish.
She added some onion to the mixture.
The ratio of the weight of fish to the weight of onion was 3 : 2
Work out the ratio of the weight of rice to the weight of onion. **(3 marks)**

**Worked solution video**

# Metric units

You can convert between metric units by multiplying or dividing by 10, 100 or 1000.

### Length

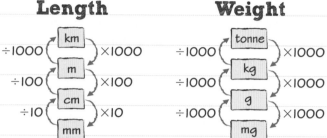

÷1000 ⟨ km ⟩ ×1000
÷100 ⟨ m ⟩ ×100
÷10 ⟨ cm ⟩ ×10
mm

### Weight

÷1000 ⟨ tonne ⟩ ×1000
÷1000 ⟨ kg ⟩ ×1000
÷1000 ⟨ g ⟩ ×1000
mg

### Volume or capacity

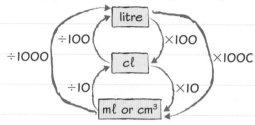

litre
÷100 ⟨ ⟩ ×100
÷1000 ⟨ cl ⟩ ×1000
÷10 ⟨ ⟩ ×10
ml or cm³

---

## Place value diagrams

You can use a place value diagram to help you multiply and divide by 10, 100 and 1000.

For more on place value look at page 1.

Hundreds | Tens | Units | • | Tenths | Hundredths | Thousandths

To multiply by 10, 100 or 1000 you move the digits 1, 2 or 3 places to the left.

| $0.35 \times 10 = 3.5$ |
| $0.35 \times 100 = 35$ |
| $0.35 \times 1000 = 350$ |

|  |  | 0 | • | 3 | 5 |
|  | 3 | • | 5 |
| 3 | 5 |
| 3 | 5 | 0 |

To divide by 10, 100 or 1000 you move the digits 1, 2 or 3 places to the right.

| $120 \div 10 = 12$ |
| $120 \div 100 = 1.2$ |
| $120 \div 1000 = 0.12$ |

| 1 | 2 | 0 |
|  | 1 | 2 |
|  | 1 | • | 2 |
|  | 0 | • | 1 | 2 |

---

## Worked example

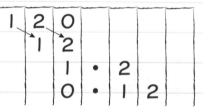

The weight of a ream of paper is 2.5 kg.
There are 500 sheets of paper in a ream.
Work out the weight, in grams, of one sheet of paper.

$2.5 \times 1000 = 2500$
$2.5\,kg = 2500\,g$
$2500 \div 500 = 5$
Each sheet of paper weighs 5 grams.

Start by converting 2.5 kg into grams.

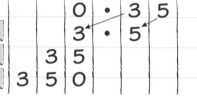

**Check it!**

A gram is a smaller unit than a kilogram so the number will be larger. ✓

The whole ream of paper weighs 2500 g and there are 500 sheets of paper, so divide 2500 by 500 to find the weight of one sheet of paper.

---

## Volume and capacity

You can convert between different metric units of volume and capacity using:

$1\,cm^3 = 1\,ml$

$1000\,cm^3 = 1\,litre$

$1\,m^3 = 1000\,litres$

You may be given one of these conversions in the exam, but you should know that
1 litre = 1000 ml.

 Worked solution video

## Now try this

1 (a) Change 3.2 m into cm. **(1 mark)**
  (b) Change 0.25 litres into ml. **(1 mark)**
  (c) Change 960 mm into cm. **(1 mark)**
  (d) Change 1700 g into kg. **(1 mark)**

2 A pile of fifty 20p coins is 8.5 cm high.
  Work out the thickness of one 20p coin.
  Give your answer in mm. **(2 marks)**

# Reverse percentages

You can use a **multiplier** to calculate a percentage increase or decrease. If you are given the final amount and are asked to find the **original amount** you can divide by the multiplier. Here are two examples.

 A sweater is reduced in price by 20% in a sale.

Original price → × 0.8 → Sale price

£50 ÷ 0.8 £40

 The average temperature increases by 5%.

Original temperature → × 1.05 → Temperature after increase

22° ÷ 1.05 23.1°

## Worked example

In a sale, normal prices are reduced by 15%
The sale price of a pair of trainers is £75.65
Work out the normal price of the trainers.
**(3 marks)**

$100\% - 15\% = 85\%$

$\frac{85}{100} = 0.85$

$75.65 \div 0.85 = 89$

The original price was £89

This question tells you the price **after** the percentage decrease, so you need to use **reverse percentages**. To find the multiplier for a 15% decrease:

1. Subtract 15% from 100%
2. Divide by 100 to convert to a multiplier.

You need to **divide** by the multiplier to find the original price.

**Check it!**
Reduce £89 by 15%:
£89 × 0.15 = £13.35
£89 − £13.35 = £75.65 ✓

## Problem solved!

Plan your answer before you start. You will need to do **two separate calculations** to find both original heights. You then need to **compare** the original heights and write a conclusion.

To find a multiplier for a percentage increase:
1. Add the percentage to 100%
2. Divide by 100 to convert to a multiplier

You need to **divide** by the multiplier to find the heights in 2013.

> You will need to use problem-solving skills throughout your exam – **be prepared!**

## Worked example

Amy and Paul measure their height each year.
This table shows their heights in 2014.

| | Height in 2014 (cm) | Percentage increase since 2013 |
|---|---|---|
| **Paul** | 154 | 10% |
| **Amy** | 150.8 | 4% |

Who was taller in 2013?
Give reasons for your answer. **(4 marks)**

Paul's original height = 154 ÷ 1.1 = 140 cm
Amy's original height = 150.8 ÷ 1.04 = 145 cm
In 2013 Amy was taller than Paul by 5 cm.

## Now try this

Worked solution video

1 Hannah buys some shoes in a sale marked '40% off'.
She pays £27 for the shoes.
What price were the shoes originally?
**(3 marks)**

2 Jared bought a house in 2010.
By 2012 his house had increased in value by 8%.
The new value of Jared's house is £237 600.
How much did Jared pay for his house? **(3 marks)**

# Growth and decay

You can use **repeated percentage change** to model problems involving growth and decay.

## Compound interest

If you leave your money in a bank account it will earn compound interest.

Saanvi invests £40000 at a compound interest rate of 3% per annum.

'Per annum' means 'per year'

This table shows the amount of interest she earns each year.

| Year | Balance (£) | Interest earned (£) |
|------|-------------|---------------------|
| 1 | 40000 | 1200 |
| 2 | 41200 | 1236 |
| 3 | 42436 | 1273.08 |
| | 43709.08 | 3709.08 |

For year 2 you have to calculate 3% of £41200

After 3 years the total interest earned will be £3709.08 and the balance will be £43709.08

## Using index notation

This table uses a multiplier to work out the balance of Saanvi's account at the end of each year.

| Year | Balance (£) |
|------|-------------|
| 1 | 40000 × 1.03 = 41200 |
| 2 | 41200 × 1.03 = 42436 |
| 3 | 42436 × 1.03 = 43709.08 |

You can use indices to work out the final balance after 3 years more easily.

Balance after 3 years

$$= £40000 × 1.03 × 1.03 × 1.03$$
$$= £40000 × 1.03^3$$
$$= £43709.08$$

For a reminder about working with indices look at pages 8 and 9.

For a reminder about working with indices look at pages 8 and 9.

---

## Worked example

At the start of an experiment a petri dish contains 5000 cells. The number of cells in the petri dish increases by 20% each day.

Calculate the number of cells in the petri dish at the end of 4 days. **(2 marks)**

$5000 × 1.2^4 = 10368$

You can use this rule to calculate a repeated percentage change.

Final amount = (starting amount) × (multiplier)$^n$

$n$ is the number of times the change is made. In this example, $n$ is the number of days.

---

## Repeated decrease

If a question involves **depreciation** or **decay** you might need to work out a repeated percentage decrease.

This car **depreciates** in value by 8% each year.

The multiplier for an 8% decrease is 0.92

After 3 years the car is worth:

£15000 × 0.92 × 0.92 × 0.92
$= £15000 × 0.92^3 = £11680.32$

## Now try this

Worked solution video

Amir invests £5000 in a savings account. He is paid 3% per annum compound interest.

(a) How much will Amir have in his savings account after 2 years? **(2 marks)**

(b) Amir needs £5600 to buy a car. Calculate the number of years Amir would need to leave his money in the account to save up this amount. **(2 marks)**

# Speed

This is the **formula triangle** for **speed**.

Average speed $= \dfrac{\text{total distance travelled}}{\text{total time taken}}$

$\text{Time} = \dfrac{\text{distance}}{\text{average speed}}$  **LEARN IT!**

Distance $=$ average speed $\times$ time

### Using a formula triangle

Cover up the quantity you want to find with your finger.

The position of the other two quantities tells you the formula.

$T = \dfrac{D}{S}$    $S = \dfrac{D}{T}$    $D = S \times T$

## Units

The most common units of speed are

- metres per second: m/s
- kilometres per hour: km/h
- miles per hour: mph.

The units in your answer will depend on the units you use in the formula.

When distance is measured in km and time is measured in hours, speed will be measured in km/h.

## Minutes and hours

For questions on speed, you need to be able to convert between minutes and hours.

| Hours | Minutes |
|---|---|
| $\frac{1}{2}$ | 30 |
| $\frac{3}{4}$ | 45 |
| $2\frac{1}{4}$ | 135 |

You can also write this as '2 hours and 15 minutes'.
$60 + 60 + 15 = 135$

## Worked example

A plane travels at a constant speed of 600 km/h for 45 minutes.
How far has it travelled? **(3 marks)**

45 minutes $= \dfrac{45}{60}$ hour $= \dfrac{3}{4}$ hour

$D = S \times T$

$= 600 \times \dfrac{3}{4} = \dfrac{600 \times 3}{4} = \dfrac{1800}{4} = 450$

The plane has travelled 450 km.

### Speed questions

☑ Draw a formula triangle.
☑ Make sure the units match.
☑ Give units with your answer.

Make sure that the units match. Speed is given in km/h, so convert the time into hours by dividing by 60. The units of distance will be km.

## Now try this

1 Bradley cycled 170 km at an average speed of 40 kilometres per hour.
How long did it take him?
Give your answer in hours and minutes. **(3 marks)**

2 Simon drives from Newcastle to Oxford.
His average speed is 56 mph.
The journey takes 4 hours 45 minutes.
How far did he drive? **(3 marks)**

# Density

The density of a material is its mass per unit volume.

This is the formula triangle for density.

**LEARN IT!**

$$\text{Density} = \frac{\text{mass}}{\text{volume}}$$

$$\text{Volume} = \frac{\text{mass}}{\text{density}}$$

$$\text{Mass} = \text{density} \times \text{volume}$$

## Worked example

15 cm²

8 cm

The diagram shows a solid hexagonal prism.
The area of the cross-section of the prism is 15 cm².
The length of the prism is 8 cm.
The prism is made from wood with a density of 0.8 grams per cm³.
Work out the mass of the prism. **(4 marks)**

Volume of prism
$$= \text{area of cross-section} \times \text{length}$$
$$= 15 \times 8$$
$$= 120 \text{ cm}^3$$
$$M = D \times V$$
$$= 0.8 \times 120$$
$$= 96$$

The mass of the prism is 96 g.

Revise volumes of prisms on page 85.

### Units

The most common units of density are
- grams per cubic centimetre: g/cm³
- kilograms per cubic metre: kg/m³.

Mass = density × volume

You are given the density so you need to work out the volume of the prism.

You need to know the formula for the volume of a prism.

The density is in grams per cm³ and the volume is in cm³ so the mass will be in grams.

## Worked example

An iron bar has a volume of 1.2 m³ and a mass of 9444 kg. Calculate the density of iron. **(2 marks)**

$$D = \frac{M}{V} = \frac{9444}{1.2} = 7870 \text{ kg/m}^3$$

Volume is in m³ and mass is in kg so density will be in kg/m³.

## Now try this

**1** A metal rod has a density of 7.8 g/cm³ and a mass of 300 g. Circle the correct calculation for the volume of the rod in cm³.

7.8 × 300     7.8 ÷ 300     300 ÷ 7.8     **(1 mark)**

**2** The density of copper is 8.92 g/cm³.
The density of silver is 10.49 g/cm³.
20 cm³ of copper and 5 cm³ of silver are mixed together to make a new kind of metal.
Work out the density of the new metal. **(4 marks)**

**Worked solution video**

**Plan** your answer:

1. Work out the masses of each metal.
2. Add the masses together.
3. You know the total mass and the total volume of the new metal, so calculate the density.

# Other compound measures

Compound measures are made up of two or more other measurements. **Speed** is a compound measure because it is calculated using distance **and** time.

## Pressure

Pressure is a measure of the force applied over a given area. The most common units of pressure are Newtons per square centimetre ($N/cm^2$) and Newtons per square metre ($N/m^2$).

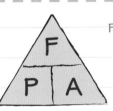

$$Pressure = \frac{Force}{Area}$$

**LEARN IT!**

$$Area = \frac{Force}{Pressure}$$

$$Force = Pressure \times Area$$

### Worked example

At a depth of 15 m, water has a pressure of $14.7\,N/cm^2$. Calculate the force applied to a diving mask with a surface area of $360\,cm^2$. **(2 marks)**

Force = Pressure × Area

= 14.7 × 360 = 5292 N

### Rates

If the bottom unit in a compound measure is **time**, then it is a **rate**.

$$Speed = \frac{Distance}{Time} \qquad \frac{Rate}{of\ climb} = \frac{Height}{Time}$$

$$\frac{Rate}{of\ flow} = \frac{Volume}{Time} \qquad \frac{Rate}{of\ pay} = \frac{Salary}{Time}$$

### Problem solved!

You need to use lots of different maths skills to solve this problem. You need to know how to calculate the volume of a cuboid (revise this on page 84) and how to convert from $cm^3$ to litres (for a reminder look at page 62). It's a good idea to **plan your strategy** before you start:

1. Work out the volume of the fishtank in litres.
2. Divide by 2 to get half the capacity.
3. Divide by the rate of flow to get the time taken in minutes.

You will need to use problem-solving skills throughout your exam – **be prepared!**

### Worked example

This fishtank can be modelled as a cuboid.

40 cm

120 cm   90 cm

Aaron fills the fishtank with water at a rate of 12 litres per minute. How long will it take for the fishtank to be half full? **(5 marks)**

Volume = 120 × 90 × 40

= 432 000 $cm^3$ = 432 litres

432 ÷ 2 = 216      216 ÷ 12 = 18

It will take 18 minutes to half fill the tank.

### Now try this

1 The average fuel consumption of a car is measured in kilometres per litre (km/l). A car travels 249 km and uses 15 litres of petrol. What is its average fuel consumption? **(2 marks)**

Look at the units to work out what calculation to do.

2 A large tank holds 1680 litres of water. The tank can be filled from a hot tap or from a cold tap. The cold tap on its own takes 4 minutes to fill the tank. The hot tap on its own takes 6 minutes to fill the tank. Evie turns both taps on at the same time. How long does the tank take to fill? **(3 marks)**

# Proportion

Two quantities are in **direct proportion** when both quantities increase at the same rate.

Number of tickets bought    Total cost

3      £135

×3      ×3

9      £405

Two quantities are in **inverse proportion** when one quantity increases at the same rate as the other quantity decreases.

Average speed    Time taken

40 km/h    2 hours

×2      ÷2

80 km/h    1 hour

---

## Worked example

Suresh buys 4 picture frames for a total cost of £11.40.
Work out the cost of 7 of these picture frames. **(2 marks)**

Cost of 1 frame = $\frac{£11.40}{4}$ = £2.85

Cost of 7 frames = £2.85 × 7 = £19.95

Calculate the cost of 1 picture frame first. Then multiply the cost of 1 frame by 7 to work out the cost of 7 frames. When you are working with money you should:

- do all your calculations in either £ or p
- write £ or p in your answer, but not both
- write answers in £ to 2 decimal places.

---

## Divide or multiply?

6 people can build a wall in 4 days.
How long would it take 8 people to build the same wall?

Inverse proportion problems often involve time. The more people working on a task, the quicker it will be finished.

You can solve this problem by working out how long it would take 1 person to build the wall. Use common sense to decide whether to divide or multiply.

6 × 4 = 24 so 1 person could build the wall in 24 days.

You multiply because it would take 1 person more time to build the wall.

24 ÷ 8 = 3 so 8 people could build the wall in 3 days.

You divide because it would take 8 people less time to build the wall.

---

## Now try this

**Worked solution video**

1   8 identical bottles of water cost £4.48.
    (a) Work out the cost of 5 of these bottles of water. **(2 marks)**
    (b) Work out the cost of 12 of these bottles of water. **(1 mark)**

2   It takes 8 men a total of 6 days to dig a hole.
    How long would it take 3 men to dig a hole of the same size? **(2 marks)**

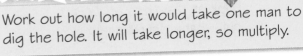

Work out how long it would take one man to dig the hole. It will take longer, so multiply.

# Proportion and graphs

You can use the symbol ∝ to show direct proportion. You can show quantities that are **directly proportional** or **inversely proportional** on a graph.

## Direct proportion facts

If $y$ is directly proportional to $x$:

✓ you can write $y \propto x$

✓ you can write an equation $y = kx$ where $k$ is a number

✓ the graph of $x$ against $y$ is a **straight line** passing **through the origin**.

$y \propto x$

## Inverse proportion facts

If $y$ is inversely proportional to $x$:

✓ You can write $y \propto \dfrac{1}{k}$    *$y$ is directly proportional to the **reciprocal** of $x$*

✓ You can write an equation $y = \dfrac{k}{x}$ where $k$ is a number

✓ The graph of $x$ against $y$ looks like a **reciprocal graph**.

$y \propto \dfrac{1}{x}$

---

## Worked example

$p$ and $q$ are inversely proportional. Circle the equation that could describe the relationship between $p$ and $q$.    **(1 mark)**

$p = 2q$    $p = q + 5$    $p = \dfrac{q}{10}$    $\boxed{p = \dfrac{2}{q}}$

> An equation for **inverse** proportionality looks like $y = \dfrac{k}{x}$ where $k$ is a number.

---

There are other ways to answer this question. The method used here is a bit like using **equivalent ratios**:

$$\div 4 \left( \begin{array}{c} 2.2 \;\; : \;\; 528 \\ ? \;\; : \;\; 132 \end{array} \right) \div 4$$

**Check:**

Your answer should make sense. If the quantities are in **direct proportion** then when one decreases the other should decrease as well.

## Worked example

Electrical appliances in your home follow this rule:

   Power (watts) ∝ Current (amps)

An electric drill uses 2.2 A and has a power of 528 W. Calculate the current used by a television with a power of 132 W.    **(2 marks)**

$528 \div 132 = 4$
$2.2 \div 4 = 0.55$
The television uses 0.55 A

---

## Now try this

This graph can be used to convert between inches and centimetres.

(a) Use the graph to convert 12.5 inches to cm.    **(1 mark)**

(b) Use the graph to convert 1 m to inches.    **(1 mark)**

(c) What evidence is there from the graph to show that inches are directly proportional to centimetres?    **(2 marks)**

# Problem-solving practice 1

About half of the questions in your Foundation GCSE exam will require you to **problem-solve, reason, interpret** or **communicate** mathematically. If you come across a tricky or unfamiliar question in your exam you can try some of these strategies:

- ✓ Sketch a diagram to see what is going on.
- ✓ Try the problem with smaller or easier numbers.
- ✓ Plan your strategy before you start.
- ✓ Write down any formulae you might be able to use.
- ✓ Use *x* or *n* to represent an unknown value.

---

 A bag contains green, red and blue counters.

20% of the counters are green.

There are three times as many red counters as blue counters.

There are 9 blue counters in the bag.

How many counters are there in the bag in total?                    (4 marks)

### Percentages page 56

There is more than one way to approach this question. One way is to start by working out how many red and blue counters there are in total. This represents 80% of all the counters in the bag.

**TOP TIP**

You can sometimes solve percentage problems by working out what 1% represents and then multiplying by 100.

---

 Suresh wants to buy a new pair of trainers.

There are three shops that sell the trainers he wants.

| Sportcentre Trainers | Footwear First Trainers | Action Sport Trainers |
|---|---|---|
| £10 plus 10 payments of £3.50 | $\frac{1}{4}$ off usual price of £80 | £40 plus VAT at 20% |

Which shop is selling the trainers the cheapest?
You must show working to justify your answer.                    (5 marks)

### Fractions, decimals and percentages page 57
### Percentage change 1 page 58

There are lots of steps in this question so make sure you keep track of your working. You need to calculate the price at each shop, then write down which shop is cheapest.

**TOP TIP**

Divide the answer space into three columns. Then the examiner can see which shop each bit of working is for.

# Problem-solving practice 2

 **3** At a sixth form college, the ratio of male to female students is 7 : 6
There are 144 female students at the college. How many students are there in total?

(2 marks)

### Ratio 1 page 60

Use the information in the question to work out what each part of the ratio represents. There are 7 + 6 = 13 parts in the ratio, so multiply your answer by 13 to work out the total number of students.

**TOP TIP**

Check your answer by dividing it in the ratio 7 : 6

---

**4** Joshua invested £2000 in a savings account for two years. The account pays 2% per annum compound interest.

At the end of the first year he put an extra £1500 into the account.

How much money will Joshua have at the end of the two years? (4 marks)

### Growth and decay page 64

Work out the interest Joshua earns on £2000, then add the extra £1500. This is the new starting amount for the second year. So in the second year, Joshua earns interest on this new amount.

**TOP TIP**

At each step of your working, make a note of what you have worked out. Your notes here could be:
- Interest earned in year 1
- Total starting amount for year 2
- Interest earned in year 2
- Total amount at end of year 2

---

 **5** This item appeared in a newspaper.

> **Cow produces 3% more milk**
> A farmer found that when his cow listened to classical music the milk it produced increased by 3%.
> This increase of 3% represented 0.72 litres of milk.

Calculate the amount of milk produced by the cow when it listened to classical music. (3 marks)

### Proportion page 68

When the cow listened to classical music, it produced 103% of the milk it produced originally. You know that 3% represents 0.72 litres. Use this information to work out what 103% represents.

**TOP TIP**

You can sometimes solve percentage problems by working out what 1% represents.

# Symmetry

## Lines of symmetry

A line of symmetry is a mirror line. One half of the shape is a mirror image of the other.

No lines of
symmetry

1 line of symmetry

### Using tracing paper

You are allowed to ask for tracing paper in your exam. You can use it to check for lines of symmetry.

If you fold a tracing of a shape in half along a line of symmetry, the two halves will match up exactly.

2 lines of symmetry

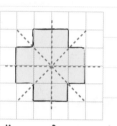

4 lines of symmetry

## Rotational symmetry

If a shape fits over itself when it is rotated then it has rotational symmetry. The order of rotational symmetry tells you how many times it fits over itself in one full turn.

No rotational
symmetry

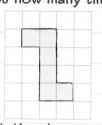

Rotational symmetry
of order 2

### Using tracing paper

You can use tracing paper to check for rotational symmetry.

Trace the shape. Rotate the tracing paper and see how many times the shape fits over itself.

This shape fits over itself twice: once at 180° and once at 360°. It has rotational symmetry of order 2.

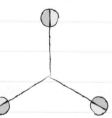

Rotational symmetry
of order 3

Rotational symmetry
of order 4

## Now try this

 (a) Fill in 5 squares of the grid below to make a shape with exactly 4 lines of symmetry.

**(1 mark)**

Worked solution video

 (b) Fill in 5 squares of the grid below to make a shape with rotational symmetry of order 2.

**(1 mark)**

# Quadrilaterals

A **quadrilateral** is any four-sided shape. The lines joining the opposite corners of a quadrilateral are called **diagonals**.

## Square

*Bisect* means 'cut in half exactly'.

All sides equal. ✓
All angles are 90°. ✓
Diagonals are equal and bisect each other at 90°. ✓
4 lines of symmetry. ✓
Rotational symmetry of order 4. ✓

## Rectangle

Opposite sides equal. ✓
All angles are 90°. ✓
Diagonals are equal and bisect each other. ✓
2 lines of symmetry. ✓
Rotational symmetry of order 2. ✓

## Parallelogram

Opposite sides parallel and equal. ✓
Opposite angles equal. ✓
Diagonals bisect each other. ✓
No lines of symmetry. ✓
Rotational symmetry of order 2. ✓

## Trapezium

One pair of opposite sides parallel. ✓

## Kite

Two pairs of adjacent sides equal. ✓
One pair of opposite angles equal. ✓
Diagonals cross at 90°. ✓
I line of symmetry. ✓

## Rhombus

All sides equal. ✓
Opposite sides parallel. ✓
Opposite angles equal. ✓
2 lines of symmetry. ✓
Diagonals bisect each other at 90°. ✓
Rotational symmetry of order 2. ✓

---

## Now try this

Join the name of each 2D shape to its correct description with a straight line.
The first one has been done for you.

**(5 marks)**

| Square | | Diagonals cross at 90° to each other. Rotational symmetry of order 2. |
| Rectangle | | Diagonals bisect each other but are not at 90°. No lines of symmetry. |
| Rhombus | | One pair of parallel sides. No rotational symmetry. |
| Kite | | All angles 90°. Diagonals cross at 90° to each other. |
| Trapezium | | Diagonals cross at 90° to each other. One line of symmetry. |
| Parallelogram | | Diagonals bisect each other but are not at 90°. Two lines of symmetry. |

# Angles 1

## Types of angle

You need to know the names of the different types of angles.

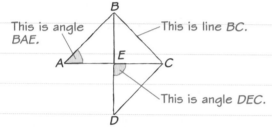

**Acute** angle:
Less than 90°

**Right** angle:
90°

**Obtuse** angle:
Between 90° and 180°

**Reflex** angle:
More than 180°

You can use these angle types to help you estimate the size of angles.

To revise measuring and drawing angles, have a look at page 96.

## Naming angles

You can use letters to name angles.

This is angle BAE.

This is line BC.

This is angle DEC.

Angles are named using the three letters of the lines that make the angle. The angle is always at the middle letter.

## Special triangles

Here are three special types of triangle.

**Isosceles**
Two equal sides
Two equal angles

**Equilateral**
Three equal sides
All angles 60°

**Right-angled**
One angle 90°

In a **scalene** triangle none of the sides or angles are equal.

## Angle facts

You can use these angle facts to work out missing angles.

**1** Angles on a straight line add up to 180°.

$a + b = 180°$

**2** Angles around a point add up to 360°.

$c + d + e = 360°$

**3** Vertically opposite angles are equal.

$f = g$

$h = i$

(a) Mark with a letter *R* a reflex angle. **(1 mark)**

(b) Mark with a letter *O* an obtuse angle. **(1 mark)**

(c) Work out the size of angle *p*. Give a reason for your answer. **(2 marks)**

$p + 52° = 180°$
$180 - 52 = 128$
$p = 128°$

Everything in red is part of the answer.

Angles on a straight line add up to 180°.

### Now try this

Work out the sizes of the angles marked with letters in this diagram.
Give reasons for your answers. **(3 marks)**

Worked solution video

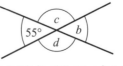

You could start by using angle fact 1 or angle fact 3 from the blue box above.

# Angles 2

## Triangles and quadrilaterals

These are useful facts for triangles and quadrilaterals.

**1** $c + d + e = 180°$

Angles in a triangle add up to 180°.

**2** $f + g + h + i = 360°$

Angles in a quadrilateral add up to 360°.

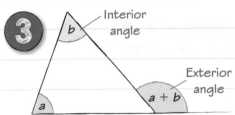

**3** The exterior angle of a triangle is equal to the sum of the interior angles at the other two vertices.

**4** The opposite angles of a parallelogram are equal.

## Parallel and perpendicular lines

**Perpendicular** lines meet at 90°.

Lines that remain the same distance apart are **parallel**.
Parallel lines are marked with arrows.
You need to remember these angle facts about parallel lines and their correct names.

 **1**

**Corresponding** angles are equal.

 **2**

**Alternate** angles are equal.

 **3**

$a + b = 180°$

**Co-interior** or **allied** angles add up to 180°.

### Worked example

Work out the value of the angle marked $x$. Show clearly, giving reasons, how you work out your answer. **(3 marks)**

Angle $CGH = 118°$
Corresponding angles are equal.
$180 - 118 = 62$
Angle $x = 62°$
Angles on a straight line add up to 180°.

### Now try this

(a) Write down the size of angle $x$. Give a reason for your answer. **(2 marks)**
(b) Work out the size of angle $y$. **(2 marks)**

Write any angles you have worked out on the diagram.

# Solving angle problems

When you are solving angle problems you need to give reasons for each step of your working. Look at pages 74 and 75 to see the reasons you can use to solve angle problems.

## Worked example

This diagram shows three angles at a point.

Work out the missing angle in terms of $p$.
Give your answer in its simplest form.
**(2 marks)**

Angles around a point add up to 360°.
$360 - 5p - 3p = 360 - 8p$

There is more about simplifying algebraic expressions like this on page 23.

## Worked example

Work out the value of $m$.
Show clearly, giving reasons, how you work out your answer. **(3 marks)**

$360 - 120 - 80 - 55 = 105$
Angles in a quadrilateral add up to 360°.
$180 - 105 = 75$
$\qquad m = 75°$
Angles on a straight line add up to 180°.

## Worked example

The arrows tell you which lines are parallel.
The dashes tell you which lines are equal.

Everything in red is part of the answer.

Angle $BDE = 74°$
Corresponding angles are equal.
Angle $DBE = 74°$
Base angles of an isosceles triangle are equal.
$180 - 74 - 74 = 32$
$\qquad x = 32°$
Angles in a triangle add up to 180°.

Work out the value of $x$.

Show clearly, giving reasons, how you work out your answer. **(4 marks)**

## Now try this

Remember to give reasons for each step of your working.

Worked solution video

**1** Work out the size of:
(a) Angle $GCD$
**(1 mark)**
(b) Angle $FCG$
**(2 marks)**

**2** In the diagram $AB$ is parallel to $CD$.

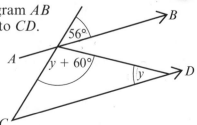

Work out the size of angle $y$.
You must show your working. **(4 marks)**

# Angles in polygons

Polygon questions are all about interior and exterior angles.

Exterior angle

Interior angle

Use these formulae for a polygon with $n$ sides.

Sum of interior angles = $180° \times (n - 2)$

Sum of exterior angles = $360°$

This diagram shows part of a **regular** polygon with 30 sides.

168°  12°

Each exterior angle is $\frac{360°}{30} = 12°$

Each interior angle is $180° - 12° = 168°$

Don't try to draw a 30-sided polygon!

If there's no diagram given in a polygon question, you probably don't need to draw one.

## Regular polygons

In a regular polygon all the sides are equal and all the angles are equal.

If a regular polygon has $n$ sides then each exterior angle is $\frac{360°}{n}$

Regular pentagon $n = 5$ — $\frac{360°}{5}$ — 72°

Regular hexagon $n = 6$ — 60° $\frac{360°}{6}$

Regular octagon $n = 8$ — $\frac{360°}{8}$ — 45°

You can use the fact that the angles on a straight line add up to 180° to work out the size of one of the interior angles.

## Worked example

The diagram shows part of a regular polygon.

Work out the number of sides in the polygon.

156°

**(3 marks)**

Exterior angle = $180° - 156° = 24°$

$\frac{360°}{n}$ so $n = \frac{360°}{24°} = 15$

The polygon has 15 sides.

## Problem solved!

It's usually easier to work with **exterior angles** than interior angles. Use the fact that angles on a straight line add up to 180° to calculate the size of one exterior angle.

In a regular $n$-sided polygon, each exterior angle is $\frac{360°}{n}$. You can use this fact to write an equation and solve it to find $n$.

You will need to use problem-solving skills throughout your exam – **be prepared!**

## Now try this

$JK$, $KL$ and $LM$ are three sides of a 20-sided regular polygon.

$JK$ and $ML$ are extended to meet at $T$.

Work out the size of angle $KTL$, marked $x$ on the diagram. **(4 marks)**

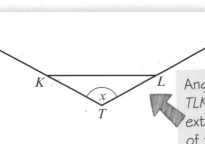

Angles $TKL$ and $TLK$ are both external angles of the 20-sided polygon.

# Time and timetables

## Time

 You can write time using the 12-hour clock or the 24-hour clock.

| 12-hour clock | 24-hour clock |
|---|---|
| 8.15 am | 08:15 |
| 4.50 pm | 16:50 |
| 12.00 midday | 12:00 |
| 12.00 midnight | 00:00 |

 Remember there are 60 minutes in an hour.

| Hours | Minutes |
|---|---|
| $\frac{1}{2}$ | 30 |
| $\frac{3}{4}$ | 45 |
| $2\frac{1}{4}$ | 135 |

You can also write this as '2 hours and 15 minutes'.
60 + 60 + 15 = 135

## Timetables

This table shows part of a bus timetable.

Timetables usually give times using the 24-hour clock.

| Crook | 08:15 | 09:15 | 10:45 | 11:15 |
|---|---|---|---|---|
| Prudhoe | 08:28 | 09:28 | 10:58 | 11:28 |
| Hexham | 08:45 | 09:45 | 11:15 | 11:45 |
| Alton | 09:00 | 10:00 | 11:30 | 12:00 |

This bus leaves Crook at 10:45 and arrives in Hexham at 11:15

This bus leaves Prudhoe at 11:28 and arrives in Alton at 12:00
11:28 to 11:30 = 2 minutes
11:30 to 12:00 = 30 minutes
The journey time is 32 minutes.

## Worked example

Olivia is visiting her sister in Exeter. She leaves home at 11.30 am. She writes down how long each section of her journey takes.

Drive to train station   15 minutes
Wait for train           25 minutes
Train journey            $1\frac{1}{2}$ hours

At what time does Olivia arrive in Exeter?

11.30 am → 11.45 am → 12.10 pm → 1.40 pm
Olivia arrives at 1.40 pm.

Add on each time in steps. Write down each new time.

+40 min

11.30 am +15 min 11.45 am +25 min 12.10 pm

+1 $\frac{1}{2}$ hours

12.10 pm +1 hour 1.10 pm +30 min 1.40 pm

## Now try this

The clocks show the time in London and San Francisco.
It is afternoon in London.

(a) Write down the time in London using the 24-hour clock.   **(1 mark)**

It is morning in San Francisco when it is afternoon in London.
The time in London is ahead of the time in San Francisco.

(b) How many hours is London ahead of San Francisco?   **(1 mark)**

LONDON       SAN FRANCISCO

# Reading scales

Here are three things you need to watch out for when reading scales.

**1** You need to be able to read scales and number lines. Begin by working out what each division on a scale represents.

There are 5 divisions between 100 g and 150 g.
Each division represents 10 g.
The scale reads 120 g.

**2** Not all divisions represent 1 unit or 10 units.

There are 5 divisions between 3.9 m and 4 m.
0.1 ÷ 5 = 0.02
Each division represents 0.02 m.
This wall is 3.92 m long.

## Worked example

The diagram below shows four identical cubes and four identical tetrahedrons.

Work out the weight in kg of one tetrahedron.
**(3 marks)**

4.5 ÷ 3 = 1.5 so 1 cube = 1.5 kg
2.5 − 1.5 = 1
1 ÷ 4 = 0.25 so 1 tetrahedron = 0.25 kg

**3** Sometimes you have to **estimate** the reading on a scale.

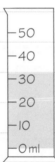

The water doesn't come up to an exact mark but you can make an estimate.
The water is closer to 30 ml than 40 ml.
32 ml would be a good estimate.

## Problem solved!

Plan your strategy before you start.
1 Divide the total on the first scale by 3 to work out the weight of one cube.
2 Subtract that from the reading on the second scale.
3 Divide the remainder by 4 to work out the weight of one tetrahedron.

Make sure you write down your working so you can show the strategy you used.

> You will need to use problem-solving skills throughout your exam – **be prepared!**

## Now try this

This scale has 3 identical spheres on it.

Work out the weight in grams of one sphere.
**(2 marks)**

# Perimeter and area

## Perimeter

**Perimeter** is the distance around the edge of a shape. You can work out the perimeter of a shape by adding up the lengths of the sides.

Perimeter = 3 cm + 5 cm + 6 cm
$$= 14 \text{ cm}$$

You might need to measure a shape to find the perimeter.

See page 97 for help on measuring lines.

See page 97 for help on measuring lines.

### Worked example

Work out the perimeter of this parallelogram. **(2 marks)**

3 + 8 + 3 + 8 = 22
Perimeter = 22 m

> Everything in red is part of the answer.

Work out the missing lengths first. The opposite sides of a parallelogram are equal so you can fill in these lengths on the diagram.

## Area

You can work out the **area** of a shape drawn on squared paper by counting the squares.

This area is 1 cm²
You say 'one centimetre squared'
or
'one square centimetre'

### Worked example

This shape is drawn on cm squared paper.
(a) Work out the perimeter of the shape. **(1 mark)**
18 cm
(b) Work out the area of the shape. **(1 mark)**
12 cm²

## Estimating

You might need to estimate the area of a shape drawn on cm squared paper. Count 1 cm² for every whole square and $\frac{1}{2}$ cm² for every part square.

Here there are 10 whole squares and 6 part squares. A good estimate is 13 cm².

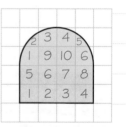

### Now try this

Estimate the area of this oval shape. Each square represents 1 cm². **(2 marks)**

> Count 1 cm² for each whole square and $\frac{1}{2}$ cm² for each part square.

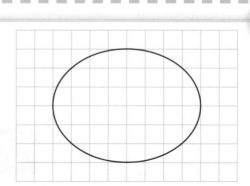

# Area formulae

You will not be given any of the formulae on this page in your exam. Make sure you **learn** them and know how to use them.

### Rectangle

Area = length × width

$A = lw$

### Parallelogram

Area = base × vertical height

$A = bh$

### Triangle

Area = $\dfrac{\text{base} \times \text{vertical height}}{2}$

$A = \frac{1}{2}bh$ — $h$ is the vertical height.

### Trapezium

 **LEARN IT!**

Area = $\frac{1}{2}\left(\begin{array}{c}\text{sum of the two}\\ \text{parallel sides}\end{array}\right) \times \begin{array}{c}\text{vertical}\\ \text{height}\end{array}$

$A = \frac{1}{2}(a + b)h$

### Area checklist

✓ Make sure lengths are all in the same units.

✓ Remember to give units with your answer.    Lengths in cm then area in cm². Lengths in m then area in m².

---

## Worked example

Work out the area of this triangle. **(2 marks)**

$A = \frac{1}{2}bh$

$= \frac{1}{2} \times 9 \times 6$

$= 27\,\text{cm}^2$

Multiplying by $\frac{1}{2}$ is the same as dividing by 2.

## Worked example

The diagram shows a rug in the shape of a trapezium. Work out the area of the rug. **(3 marks)**

$A = \frac{1}{2}(a + b)h$

$= \frac{1}{2} \times (2 + 5) \times 3$

$= \frac{1}{2} \times 7 \times 3$

$= 10.5\,\text{m}^2$

**Learn** the formula for the area of a trapezium. You should always **write down** the formula before you substitute any values.

---

Worked solution video

## Now try this

Here is a diagram of a parallelogram and a rectangle. The parallelogram has the same area as the rectangle. Work out the length $x$. **(4 marks)**

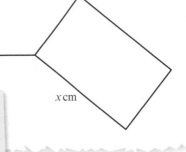

Plan your answer before you start writing. You will need to work out the area of the parallelogram first. You can write any lengths you find on your diagram.

# Solving area problems

You can calculate areas and perimeters of harder shapes by splitting them into parts.
You might need to draw some extra lines on your diagram and add or subtract areas.

Area = rectangle + triangle        Area = triangle − rectangle        Area = $\dfrac{\text{large}}{\text{rectangle}}$ − $\dfrac{\text{small}}{\text{rectangle}}$

## Worked example

The diagram shows a garden bed.

Adrian wants to cover the bed with grass seed.
A packet of grass seed will cover 10 m².

(a) How many packets of grass seed does
Adrian need to buy?            **(3 marks)**

Area = 6 × 4.5 + 4 × 3
     = 27 + 12
     = 39 m²

Adrian needs to buy 4 packets of grass seed.

Adrian also wants to build a fence around the edge of the garden bed.

(b) Calculate the total length of Adrian's
fence.                          **(1 mark)**

6 + 4.5 + 3 + 4 + 3 + 8.5 = 29 m

> Everything in red is part of the answer.

### Golden rule

Always work out any missing lengths before calculating the area or perimeter.

Draw a dotted line to divide the garden bed into two rectangles.

You have to use the information in the question to work out the missing lengths.

8.5 m − 4.5 m = 4 m

6 m − 3 m = 3 m

Write these lengths on your diagram.

Make sure you answer the question that has been asked.

For part (a) you need to say how many packets of grass seed Adrian needs to buy.

Don't measure the distances. Unless you are told otherwise, all the diagrams in your exam are not drawn accurately.

## Now try this

Here is a shape made up of rectangles.
(a) Work out the distance marked $x$ on the diagram.
                                **(1 mark)**
(b) Work out the distance marked $y$ on the diagram.
                                **(1 mark)**
(c) Work out the area of this shape.    **(3 marks)**
(d) Work out the perimeter of this shape.  **(2 marks)**

# 3D shapes

You need to learn the names of these 3D shapes.

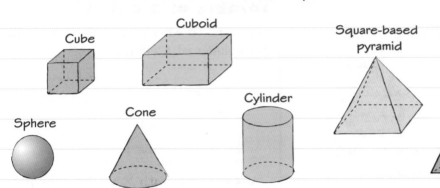

Cube   Cuboid   Square-based pyramid   Triangular prism   Cylinder   Cone   Sphere

---

## Faces, edges and vertices

This square-based pyramid has 5 faces, 8 edges and 5 vertices.

The plural of vertex is vertices.

— Face
— Vertex
— Edge

> You need to **learn** the mathematical names of 3D solids. You can sketch the hidden edges to help you work out the answer to part (b).

## Worked example

(a) Write down the name of each 3D shape.

(i)   (ii)

Cuboid   Tetrahedron

(b) Write down the number of edges on shape (i).

12

> Everything in red is part of the answer.

---

## Surface area of a cuboid

You can find the surface area of a 3D solid by adding together the areas of its faces. A cuboid has **six** faces. You can use the fact that **opposite faces** are **equal** to simplify your working.

Surface area  $= 2A + 2B + 2C$

$= 2 \times (6 \times 3) + 2 \times (6 \times 4) + 2 \times (4 \times 3)$

$= 2 \times 18 + 2 \times 24 + 2 \times 12 = 108\,\text{cm}^2$

3 cm, 4 cm, 6 cm

---

## Now try this

The diagram shows a solid block of wood in the shape of a cuboid.

(a) Work out the total surface area of the cuboid. **(2 marks)**

Karl has to paint all 6 faces of 50 cuboids.

A can of spray paint covers an area of $3000\,\text{cm}^2$.

(b) How many cans of paint must Karl buy to paint all the faces? Show all your working. **(3 marks)**

5 cm   5 cm   10 cm

> Don't just write down a number or estimate. You must show all your working.

# Volumes of cuboids

The **volume** of a 3D shape is the amount of space it takes up.

The most common units of volume are cm³ or m³.

Volume = 1 cm³          Volume = 10 cm³

This shape is made from ten 1 cm³ cubes.

## Volume of a cuboid

You need to remember this formula for the volume of a **cuboid**.

height
width
length

Volume = length × width × height

---

## Worked example

The diagram shows a wooden planting box.

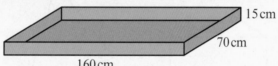

15 cm
70 cm
160 cm

A 50 litre bag of compost costs £3.99.
1 litre = 1000 cm³
How much will it cost to fill this planting box?          **(4 marks)**

Volume = 160 × 70 × 15
       = 168 000 cm³
       = 168 litres

4 bags of compost are needed.
4 × £3.99 = £15.96

## Problem solved!

There are a few steps to this problem so it's a good idea to plan your answer:
1. Work out the volume of the planting box.
2. Convert from cm³ to litres by dividing by 1000.
3. Work out how many bags of compost are needed.
4. Work out how much the compost will cost. Remember that you can only buy a **whole number** of bags of compost.
3 × 50 litres = 150 litres
4 × 50 litres = 200 litres
You need 168 litres of compost, so you need to buy 4 bags.

> You will need to use problem-solving skills throughout your exam – **be prepared!**

---

## Worked example

2.0 m
1.5 m
3.2 m

Work out the volume of this cuboid.          **(2 marks)**

Volume = length × width × height
       = 3.2 × 1.5 × 2.0
       = 9.6 m³

> The units of length are m so the units of volume will be m³.

## Now try this

This cuboid has a volume of 390 cm².

h cm
15 cm
4 cm

Work out the height, h cm, of the cuboid.          **(3 marks)**

15 × 4 × h = 390

# Prisms

## Volume

A prism is a 3D solid with a **constant cross-section**. Use this formula to calculate the volume of a prism.

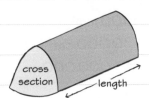

cross section — length

Volume = area of cross-section × length

**LEARN IT!**

### Worked example

The diagram shows a prism. The cross-section is a trapezium. Work out the volume of the prism.

5 cm    6 cm    7 cm    10 cm

**(3 marks)**

Area of cross-section (trapezium)
$$= \frac{1}{2} \times (6 + 10) \times 5 = 40\,cm^2$$
Volume of prism $= 40 \times 7 = 280\,cm^3$

## Surface area

To work out the surface area of a 3D shape you need to add together the areas of all the faces.

It's a good idea to sketch each face with its dimensions. Remember to include the faces that you can't see.

4 m    5 m    8 m    3 m

40 m²  5 m    24 m²  3 m    8 m
32 m²  4 m    4 m    4 m
8 m    6 m²  3 m    6 m²  3 m

Surface area $= 40 + 32 + 24 + 6 + 6 = 108\,m^2$

### Worked example

The diagram shows a triangular prism and a cube. They both have the **same** volume. Work out the length of $x$.    **(4 marks)**

5 cm    12 cm    $x$    9 cm

Volume of cube $= 9^3 = 729\,cm^3$
Volume of prism $=$ Area of cross-section
$\qquad \times$ length
$\qquad = \frac{1}{2} \times 5 \times 12 \times x = 30x$
$\qquad 30x = 729$
$\qquad x = 24.3\,cm$

### Problem solved!

Calculate the volume of the cube, and write an expression for the volume of the prism. Set these equal to each other and solve the equation to find $x$.

> You will need to use problem-solving skills throughout your exam – **be prepared!**

> Write down what you are calculating at each stage. You can get marks for your working even if your final answer is wrong.

### Now try this

The cross-section of this prism is a right-angled triangle.
(a) Work out the volume of the prism.    **(3 marks)**
(b) Work out the total surface area of the prism.    **(3 marks)**

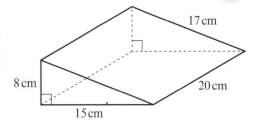

17 cm    8 cm    20 cm    15 cm

> Use this formula to work out the areas of the triangular faces:
> Area of a triangle $= \frac{1}{2} \times$ base $\times$ vertical height

85

# Units of area and volume

Converting units of area or volume is trickier than converting units of length. You need to remember your area and volume conversions for your exam.

These two squares have the same area.

So $1\,cm^2 = 100\,mm^2$.

These two cubes have the same volume.

So $1\,cm^3 = 1000\,mm^3$.

## Area conversions

$1\,cm^2 = 10^2\,mm^2 = 100\,mm^2$

$1\,m^2 = 100^2\,cm^2 = 10\,000\,cm^2$

$1\,km^2 = 1000^2\,m^2 = 1\,000\,000\,m^2$

## Volume conversions

$1\,cm^3 = 10^3\,mm^3 = 1000\,mm^3$

$1\,m^3 = 100^3\,cm^3 = 1\,000\,000\,cm^3$

$1\,litre = 1000\,cm^3$

$1\,ml = 1\,cm^3$

There is more on converting metric units on page 62.

## Worked example

Lead has a density of $11\,350\,kg/m^3$.
An antique lead model has a volume of $400\,cm^3$.
Calculate the mass of the model in kg.

**(3 marks)**

$400 \div 100^3 = 0.0004$

Volume $= 0.0004\,m^3$

Mass $=$ Density $\times$ Volume

$\quad = 11\,350 \times 0.0004$

$\quad = 4.54\,kg$

## Unit conversion checklist

✓ The multiplier for an area conversion is the length multiplier squared.

✓ The multiplier for a volume conversion is the length multiplier cubed.

| m | $m^2$ | $m^3$ |
| :---: | :---: | :---: |
| ×100 | ×100² | ×100³ |
| cm | $cm^2$ | $cm^3$ |

## Problem solved!

You need to be really careful with the **units** when you are solving any question involving measures. The units of density are given in $kg/m^3$, so you need to convert $400\,cm^3$ into $m^3$ before you calculate. You are converting to a larger unit, so **divide** by $100^3$.

For a reminder about density look at page 66.

You will need to use problem-solving skills throughout your exam – **be prepared!**

## Now try this

1 Convert
   (a) $2.3\,m^2$ into $cm^2$
   **(1 mark)**
   (b) $400\,mm^3$ into $cm^3$
   **(1 mark)**

2 Convert $0.35\,m^3$ to $mm^3$, giving your answer in standard form. **(2 marks)**

3 Jenny applies a force of $600\,N$ to the floor. The total area of her feet is $160\,cm^2$. What is the pressure, in $N/m^2$, between her and the floor if she stands on both of her feet? **(3 marks)**

# Translations

A translation is a sliding movement. You can describe a translation using a **vector**.

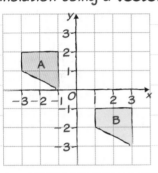

The transformation A → B is a translation by the vector.

$$\begin{pmatrix} 4 \\ -3 \end{pmatrix}$$

The top number describes the horizontal movement
- positive number = movement to the right
- negative number = movement to the left.

The bottom number describes the vertical movement
- positive number = movement up
- negative number = movement down.

Translated shapes are **congruent**.
For a reminder about congruent shapes have a look at page 110.

## Worked example

Describe fully the single transformation that will map shape P onto shape Q.  **(2 marks)**

Translation by the vector $\begin{pmatrix} 4 \\ -1 \end{pmatrix}$

To describe a translation fully you need to write the word 'translation' and the vector.

Remember to use **positive** numbers for movement to the right or up, and **negative** numbers for movement to the left or down.

You could also say 'Translation 4 squares to the right and 1 square down'. But don't use the word 'across' to describe a translation as that could mean across either direction.

For translations, the lengths of sides and the angles in the shapes do not change.

## Now try this

Here are some shapes on a grid.
(a) Translate the shaded shape A by the vector $\begin{pmatrix} -5 \\ 4 \end{pmatrix}$.

**(2 marks)**

(b) Write down the vector of the translation that maps shape B onto shape C.  **(2 marks)**

Your vector will be in the form $\begin{pmatrix} x \\ y \end{pmatrix}$ where x is the horizontal movement and y is the vertical movement. Because the shape moves **down** and to the **left** both numbers will be **negative**.

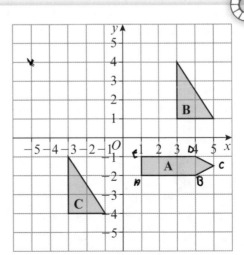

# Reflections

You can **reflect** a shape in a mirror line. To describe a reflection you need to give the **equation** of the mirror line.

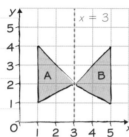

The transformation A → B is a reflection in the line $x = 3$.

Reflected shapes are **congruent**.

## Worked example

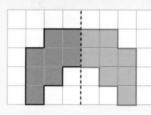

Reflect the shaded shape in the mirror line.

> Everything in red is part of the answer.

**(2 marks)**

> The mirror line is a line of symmetry.

---

## Quick reflections

It is easy to reflect shapes and check your answers using tracing paper.

Trace the original shape including the mirror line.

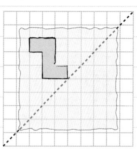

Turn the diagram so that the mirror line is vertical. Turn the tracing paper over, lining up the mirror lines.

Trace the shape in the new position.

---

## Worked example

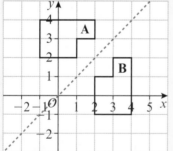

> Everything in red is part of the answer.

Describe fully the single transformation that will map shape A onto shape B. **(2 marks)**

Reflection in the line $y = x$.

> You need to write the word 'reflection' and give the equation of the mirror line.
>
> For a reminder about naming straight lines look at pages 39 and 40.

## Now try this

(a) Reflect triangle A in the $y$-axis. Label the image B. **(1 mark)**

(b) Reflect triangle B in the $x$-axis. Label the image C. **(1 mark)**

(c) Reflect triangle C in the line $y = x$. Label the image D. **(2 marks)**

(d) Describe fully the single transformation that maps triangle D onto triangle A. **(2 marks)**

# Rotations

To describe a **rotation** you need to give

- the centre of rotation
- the angle of rotation
- the direction of rotation.

The centre of rotation is often the origin O. Otherwise it is given as coordinates.

The angle of rotation is given as 90° (one quarter turn) or 180° (one half turn).

The direction of rotation is given as clockwise or anticlockwise.

**Watch out!** You don't need to give a direction for a rotation of 180°.

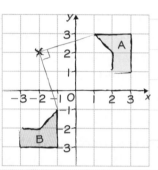

A to B: Rotation 90° clockwise about the point (−2, 2).

Give the angle in degrees. Don't say 'a quarter turn'.

You are allowed to ask for tracing paper in the exam. This makes it really easy to rotate shapes and check your answers.

Rotated shapes are **congruent**.

For a reminder about congruent shapes have a look at page 110.

---

## Worked example

Everything in red is part of the answer.

On the grid, rotate shape A 180° about (0, 1).

**(2 marks)**

Mark the centre of rotation (0, 1) with a X.

Trace the shape and put your pencil or compass on the X. Rotate the tracing paper to rotate the shape.

---

## Now try this

Describe fully the rotation that maps shape P onto:

(a) shape Q       **(3 marks)**

(b) shape R       **(3 marks)**

You need to write the word 'rotation' and give:

- the angle of turn
- the direction of the turn
- the centre of rotation.

Worked solution video

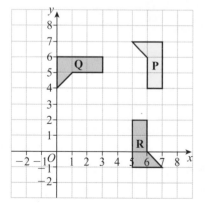

# Enlargements

To describe an enlargement you need to give the scale factor and the centre of enlargement.

The **scale factor** of an enlargement tells you how much each length is multiplied by.

$$\text{Scale factor} = \frac{\text{enlarged length}}{\text{original length}}$$

Lines drawn through corresponding points on the object (A) and image (B) meet at the **centre of enlargement**.

When the scale factor is between 0 and 1, image B is **smaller** than object A.

For enlargements, angles in shapes do not change but lengths of sides do change.

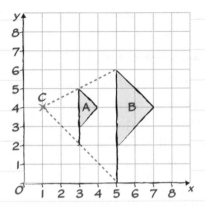

A to B: Each point on B is twice as far from C as the corresponding point on A.
Enlargement with scale factor 2, centre (1, 4)

---

## Worked example

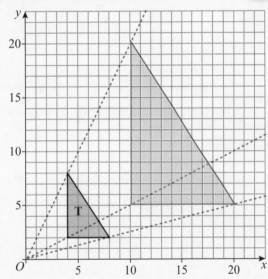

On the grid, enlarge triangle T with a scale factor of $2\frac{1}{2}$ and centre (0, 0)    **(3 marks)**

1. Draw lines from the centre of enlargement through each vertex of the triangle.

2. For each vertex, multiply the vertical and horizontal distances from the centre of enlargement by $2\frac{1}{2}$

   For the top vertex:

   Horizontal distance = $4 \times 2\frac{1}{2} = 10$

   Vertical distance = $8 \times 2\frac{1}{2} = 20$

   The corresponding vertex on the image is 10 squares horizontally and 20 squares vertically from the centre of enlargement.

3. Join up your vertices with straight lines.

**Check it!**

Each length on the image should be $2\frac{1}{2}$ times the corresponding length on the object. The image is **mathematically similar** to the object, so check it looks the same shape.

There is more on similar shapes on pages 110 and 111.

---

## Now try this

Triangle A is shown on the grid.

(a) Enlarge triangle A with a scale factor of 2 and centre of enlargement (6, 5). Label the image B.    **(2 marks)**

(b) Enlarge triangle A with a scale factor of $\frac{1}{2}$ and centre of enlargement (−7, 3). Label the image C.    **(3 marks)**

The scale factor is a **fraction** between 0 and 1 so the image will be **smaller** than the object.

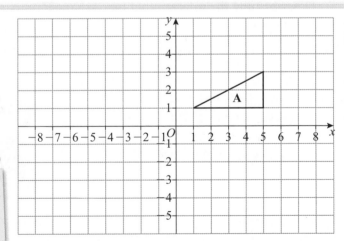

# Pythagoras' theorem

Pythagoras' theorem is a really useful rule. You can use it to find the length of a missing side in a right-angled triangle.

**LEARN IT!**

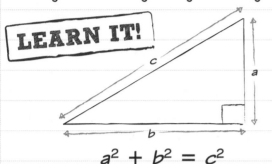

$$a^2 + b^2 = c^2$$

### Pythagoras checklist

- ✓ short$^2$ + short$^2$ = long$^2$
- ✓ Right-angled triangle.
- ✓ Lengths of two sides known.
- ✓ Length of third side missing.
- ✓ Learn this. It won't be given in the exam.

---

## Worked example

This right-angled triangle has sides $x$, 17 cm and 8 cm.

Everything in red is part of the answer.

$a^2 + b^2 = c^2$

Show that $x = 15$ cm.    **(2 marks)**

$$x^2 + 8^2 = 17^2$$
$$x^2 = 17^2 - 8^2$$
$$= 225$$
$$x = \sqrt{225} = 15 \text{ cm}$$

The question says 'show that' so you have to show **all** your working. Be careful when the missing length is one of the **shorter** sides.

1. Label the longest side of the triangle $c$.
2. Label the other two sides $a$ and $b$.
3. Write out the formula for Pythagoras' theorem.
4. Substitute any values you know into the formula.
5. Rearrange the formula and solve. Make sure you show **every step** in your working.
6. Write units in your answer.

---

Pythagoras questions come in lots of different forms. Just look for the right-angled triangle.

Flagpoles    Ramps    Ladders

### Calculator skills

Use these buttons to find squares and square roots with your calculator.

You might need to use the ⬛S⇔D key to get your answer as a decimal number.

---

## Now try this

(a) Work out the value of $y$.    **(2 marks)**

(b) Use your value of $y$ to work out the value of $z$.    **(2 marks)**

# Line segments

A **line segment** is a section of a straight line between two points. You can use Pythagoras' theorem to find the length of a line segment.

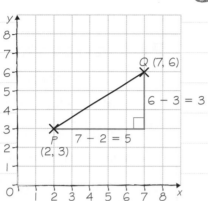

Draw a right-angled triangle with $PQ$ as its longest side.

$PQ^2 = 5^2 + 3^2$
$= 25 + 9$
$= 34$
$PQ = \sqrt{34}$
$= 5.8309...$
$= 5.83$ (2 decimal places)

For a reminder of how to find the midpoint of a line segment have a look at page 37.

---

## Worked example

Point $A$ has coordinates $(2, 5)$.
Point $B$ has coordinates $(3, -2)$.
Calculate the length of the line segment $AB$.
Give your answer correct to 2 decimal places.

**(3 marks)**

Horizontal distance $= 3 - 2 = 1$
Vertical distance $= 5 - -2 = 7$
$AB^2 = 1^2 + 7^2$
$= 50$
$AB = \sqrt{50}$
$= 7.07106... = 7.07$ (2 d.p.)

1. Sketch $x$- and $y$-axes.
2. Mark points $A$ and $B$ on your sketch.
3. Draw a right-angled triangle with $AB$ as its hypotenuse.
4. Work out the length of the two short sides of the triangle.
5. Use Pythagoras' theorem to work out the length of $AB$.
6. Round your answer to 2 decimal places.

**Watch out!**
Be really careful if any of the coordinates is negative.
The height of the triangle is
$5 - -2 = 5 + 2 = 7$
You are working out a length so you can only substitute positive numbers into Pythagoras' theorem.

---

## Now try this

**1**

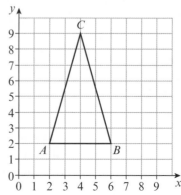

Worked solution video

Work out the length of the line segment $PQ$.
Give your answer correct to 2 decimal places. **(3 marks)**

**2** Find the perimeter of the isosceles triangle $ABC$. Give your answer to 2 decimal places.

**(4 marks)**

# Trigonometry 1

You can use the trigonometric ratios to find the size of an angle in a right-angled triangle. You need to know the lengths of two sides of the triangle.

The sides of the triangle are labelled relative to the **angle** you need to find.

> **Trigonometric ratios**   **LEARN IT!**
>
> $\sin x° = \dfrac{\text{opp}}{\text{hyp}}$ (remember this as $S^O_H$)
>
> $\cos x° = \dfrac{\text{adj}}{\text{hyp}}$ (remember this as $C^A_H$)
>
> $\tan x° = \dfrac{\text{opp}}{\text{adj}}$ (remember this as $T^O_A$)
>
> You can use $S^O_H C^A_H T^O_A$ to remember these rules for trig ratios.
> These rules only work for **right-angled** triangles.

## Worked example

Calculate the size of angle $x$. **(3 marks)**

$S^O_H \ C^A_H \ T^O_A$

$\tan x° = \dfrac{\text{opp}}{\text{adj}} = \dfrac{5}{6}$

$x° = 39.80557109° = 39.8°$ (to 3 s.f.)

Label the **hyp**otenuse first — it's the longest side.
Then label the side **adj**acent to the angle you want to work out.
Finally label the side **opp**osite the angle you want to work out.
Remember $S^O_H C^A_H T^O_A$. You know **opp** and **adj** here so use $T^O_A$.
Do **not** 'divide by tan' to get $x$ on its own. You need to use the $\tan^{-1}$ function on your calculator.

> $\tan^{-1}\left(\dfrac{5}{6}\right)$
>
> 39.80557109

Write down all the figures on your calculator display then round your answer.

## Using your calculator

To find a missing angle using trigonometry you have to use one of these functions.

$$\sin^{-1} \quad \cos^{-1} \quad \tan^{-1}$$

These are called **inverse trigonometric** functions. They are the inverse operations of sin, cos and tan.
Make sure that your calculator is in degree mode. Look for the ⬛D symbol at the top of the display.

## Now try this

Work out the size of angle $x$ in each of these triangles. Give your answers correct to 1 decimal place.

(a)

4.3 cm   6.1 cm

$x$

**(3 marks)**

(b)

11.2 cm

$x$

7.5 cm

**(3 marks)**

(c)

$x$

84 mm   127 mm

**(3 marks)**

93

# Trigonometry 2

You can use the trigonometric ratios to find the length of a missing side in a right-angled triangle. You need to know the length of another side and the size of one of the acute angles.

## Worked example

Calculate the length of side *a*.   **(3 marks)**

hyp
10 cm

*a* opp

40°

adj

S͟O͟✓ CͣA͟✓ TͦOͦ
  H    H    A

$$\sin x° = \frac{opp}{hyp}$$

$$\sin 40° = \frac{a}{10}$$

$$a = 10 × \sin 40°$$

$$= 6.42787...$$

$$= 6.43 \text{ cm (to 3 s.f.)}$$

Label the sides of the triangle relative to the 40° angle. Write $S^O{}_H C^A{}_H T^O{}_A$ and tick the pieces of information you have. You need to use $S^O{}_H$ here.
Write the values you know in the rule and replace **opp** with *a*. You can solve this equation to find the value of *a*.

Write down at least four figures from the calculator display before giving your final answer correct to 3 significant figures.

Check:
Side *a* must be shorter than the hypotenuse. 6.43 cm looks about right. ✓

## Angles of elevation and depression

Some trigonometry questions will involve angles of elevation and depression.

*d* — The angle of depression of the ship from the house.

*e* — The angle of elevation of the house from the ship.

Angles of elevation and depression are always measured from the horizontal.

In this diagram, *d* = *e* because they are alternate angles.

## Now try this

Work out the length of side *a* in each of these triangles. Give your answers correct to 1 decimal place.

In part (c) *a* is the hypotenuse. It will be on the bottom of the fraction when you substitute, so be careful with your calculation.

(a)
5.6 cm
38°
*a*

**(3 marks)**

(b)
*a*
73°
18.5 cm

**(3 marks)**

(c)
*a*
12.4 cm
52°

**(3 marks)**

# Solving trigonometry problems

You might need to answer questions involving sin, cos and tan **without a calculator**. Make sure you are confident with pages 93 and 94 before having a look at this page.

**1** This triangle shows you the values of sin, cos and tan for 30° and 60°.

$$\sin 30° = \frac{1}{2} \qquad \sin 60° = \frac{\sqrt{3}}{2}$$

$$\cos 30° = \frac{\sqrt{3}}{2} \qquad \cos 60° = \frac{1}{2}$$

$$\tan 30° = \frac{1}{\sqrt{3}} \qquad \tan 60° = \sqrt{3}$$

**2** This triangle shows you the values of sin, cos and tan for 45°.

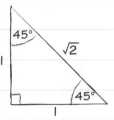

$$\sin 45° = \frac{1}{\sqrt{2}}$$

$$\cos 45° = \frac{1}{\sqrt{2}}$$

$$\tan 45° = 1$$

**LEARN IT!**

**3** You also need to know the values for 0° and 90°

| | |
|---|---|
| $\sin 0° = 0$ | $\sin 90° = 1$ |
| $\cos 0° = 1$ | $\cos 90° = 0$ |
| $\tan 0° = 0$ | |

tan 90° is **undefined**. If you enter it into a calculator you get an error.

You might get a question like this on your **non-calculator** paper, so make sure you learn the values of cos, sin and tan given above. If a triangle is **described** like this make sure you **sketch** it before doing any working.

## Worked example

Triangle *ABC* has a right angle at *B*. Angle *BAC* = 60°. *AC* = 14 cm. Calculate the length of *AB*. **(3 marks)**

$$\cos x° = \frac{adj}{hyp}$$

$$\cos 60° = \frac{AB}{14}$$

$$AB = 14 × \cos 60°$$

$$= 14 × \frac{1}{2} = 7 \, cm$$

## Worked example

Amy arranges identical triangles like this around a point to make a pattern.

Show that she can fit exactly 12 triangles around a point. **(4 marks)**

$$\sin x° = \frac{opp}{hyp} = \frac{3}{6} = \frac{1}{2}$$

Everything in red is part of the answer.

So $x° = 30°$

There are 360° around a point and 30 × 12 = 360 so she can fit exactly 12 triangles around a point.

You need to be able to answer this question **without** a calculator.

## Problem solved!

This is a non-calculator question, so you will need to use your knowledge of exact trigonometric values. Make sure you **write a conclusion** explaining how your working shows the answer.

You will need to use problem-solving skills throughout your exam – **be prepared!**

## Now try this

A vertical flagpole *AB* is supported with a wire *AC* at an angle of 60° to the ground. The base of the wire is 2.4 m from the base of the flagpole. Show that the length of the wire *AC* is 4.8 m.

**(3 marks)**

# Measuring and drawing angles

You should only measure angles in your exam if you are told that a diagram is drawn accurately.

## Measuring angles

 A protractor measures angles in degrees.

Use the scale that starts with 0 on one of the lines of the angle.

Here, use the outside scale.

Place the centre of the protractor on the point of the angle.

Line up the zero line with one line of the angle.

Read the size of the angle off the scale.

This angle is 47°.

②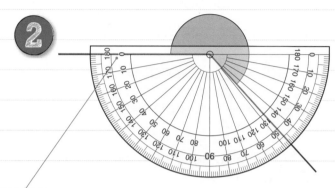

Use the scale that starts with 0 on one of the lines of the angle.
Here, use the inside scale.

To measure an angle bigger than 180° measure the smaller angle then subtract the answer from 360°.

360 – 133 = 227

The marked angle is 227°.

Estimate the size of an angle before measuring it. This lets you check that your answer is sensible.

## Drawing angles

①  Draw an angle of 23°.

Use the scale that starts with 0 on one of the lines of the angle.

Here, use the inside scale.

Use a ruler to draw one line of the angle, AB.

Place the centre of your protractor on one end of the line. The zero line needs to lie along your line.

Find 23° on the scale.
Draw a dot to mark this point.

**Worked solution video**

②

Use a ruler to join the end of the line and your dot with a straight line.

Draw in the angle curve and label your angle 23°.

### Now try this

1 Measure the size of each of these angles.
   (a)                    (b)

 Extend the lines with a pencil and ruler.

       (1 mark)              (1 mark)

2 Use a protractor to draw an angle of 124°
                                    (1 mark)

# Measuring lines

You need to be able to use a ruler to draw and measure straight lines accurately. Don't measure lines in your exam unless the question tells you that the diagram is accurate.

## Worked example

Here is a line $AB$.

Measure the length of the line $AB$.    **(1 mark)**

35 mm

> Line up the 0 mark on your ruler carefully with the start of the line at $A$.
>
> Always measure to the nearest mm.
>
> Make sure your ruler doesn't move while you're measuring the line.
>
> Always write the units with your answer.
>
> This line is 35 mm or 3.5 cm long.
>
> There is more about converting between cm and mm on page 62.

## Drawing lines checklist

☑ Check whether you are working in cm or mm.

☑ Start the line at the 0 mark on your ruler.

☑ Hold your ruler firmly.

☑ Use a sharp pencil.

☑ Draw to the nearest mm.

☑ Label the length you have drawn.

## Estimating

You can use lengths that you know to estimate other lengths.

This diagram shows a man standing at the bottom of a cliff.

The man is 3 cm tall in the drawing and the cliff is 12 cm tall. This means the cliff is 4 times as tall as the man.

A good estimate for the height of an adult male is 1.8 m.

$4 \times 1.8 = 7.2$

This means a good estimate for the height of the cliff is 7.2 m.

## Now try this

1   Measure the lengths of these lines.

    (a) ———————————

    (b) ———————————————————

    (c) —————————————

    (d) ———————————————————    **(4 marks)**

> Measure to the nearest mm.

2   The picture of a surfboard and a London Bus have been drawn accurately to the same scale. The real length of the surfboard is 5 feet. Estimate the real length of the London Bus. Show working to support your estimate.    **(3 marks)**

# Plans and elevations

Plans and elevations are 2D drawings of 3D shapes as seen from different directions.

Plan

Front        Side

Plan

The **plan** is the view from above.

Front elevation

Side elevation

This line shows a change in depth.

A **sketch** of the shape would look like this.

The **front elevation** is the view from the front.

The **side elevation** is the view from the side.

## Nets

A **net** of a 3D shape is a flat pattern that can be folded up to make the shape. This is a net of a cube.

To decide if a pattern is a net of a 3D shape imagine trying to fold it up to make the shape.

## Worked example

The diagram shows a solid shape.

Front        Side

On the grid below draw a plan, front and side elevations of the shape.

Plan    Front elevation    Side elevation

**(3 marks)**

Imagine tracing an image of the shape on each side of a box.
Unfold the box to get your plan and elevations.

You can work out the total surface area by counting the squares on the surface of the solid. Each one has an area of 1 cm². Remember to include the bottom and back of the solid.

## Now try this

This solid is made from centimetre cubes.
On separate 1 cm grids draw:
(a) the plan view of the solid          **(1 mark)**
(b) the front elevation of the solid    **(1 mark)**
(c) the side elevation of the solid.    **(1 mark)**
(d) Work out the total surface area of the solid.
                                        **(2 marks)**

Plan view

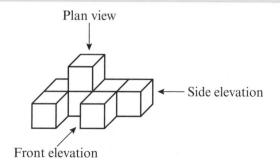

Side elevation

Front elevation

# Scale drawings and maps

This is a **scale drawing** of the Queen Mary II cruise ship.

Scale = 1 : 1000

‹————————— 34.5 cm —————————›

You can use the scale to work out the length of the actual ship.

34.5 × 1000 = 34 500

The ship is 34 500 cm or 345 m long.

## Map scales

Map scales can be written in different ways:

- 1 to 25 000
- 1 cm represents 25 000 cm
- 1 cm represents 250 m
- 4 cm represent 1 km

MAP

SCALE 1 : 25 000

---

## Worked example

The diagram shows a scale drawing of a port and a lighthouse. This diagram is drawn accurately.

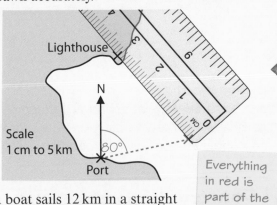

Lighthouse

N

Scale
1 cm to 5 km

80°

Port

A boat sails 12 km in a straight line on a bearing of 080°.

(a) Mark the new position of the boat with a cross. **(2 marks)**

(b) How far away is the boat from the lighthouse? Give your answer in km. **(1 mark)**

15 km

> Everything in red is part of the answer.

You can revise bearings on page 103. Start by working out how far the boat is from the port on the scale drawing.

| | ÷5 | ×12 | |
|---|---|---|---|
| Map | 1 cm | 0.2 cm | 2.4 cm |
| Real life | 5 km | 1 km | 12 km |
| | ÷5 | ×12 | |

Now place the centre of your protractor on the port with the zero line pointing North. Put a dot at 80°. Line up your ruler between the port and the dot. Draw a cross 2.4 cm from the port.

Use a ruler to measure the distance from the lighthouse to the boat. 3 cm on the drawing represents 15 km in real life.

---

**Worked solution video**

## Now try this

Read the whole question before you start. You need to give your answer in cm, so you could start by converting 20 m into cm.

The Angel of the North is a 20 m tall statue. Penny makes a model of the statue using a scale of 1 : 25

What is the height of the model of the statue?

Give your answer in centimetres. **(2 marks)**

# Constructions 1

You might be asked to construct a perpendicular line in any of these three ways.

## Worked example

Use ruler and compasses to **construct** the perpendicular to the line segment $AB$ that passes through point $P$.

> Everything in red is part of the answer.

**(2 marks)**

Use your compasses to mark two points on the line an equal distance from $P$. Keep the compasses the same and draw two arcs with their centres at these points.

## Worked example

Use ruler and compasses to **construct** the perpendicular to the line segment $AB$ that passes through point $P$.

**(2 marks)**

Use your compasses to mark two points an equal distance from $P$. Then widen your compasses and draw arcs with their centres at these two points.

## Worked example

Use ruler and compasses to **construct** the perpendicular bisector of the line $AB$.

**(2 marks)**

Use your compasses to draw intersecting arcs with centres at $A$ and $B$. Remember that to get full marks you have to be **accurate** and show **all** your construction lines.

### Constructions checklist

- ✓ Use good compasses with stiff arms.
- ✓ Use a sharp pencil.
- ✓ Use a transparent ruler.
- ✓ Mark any angles.
- ✓ Label any lengths.
- ✓ Show all construction lines.

## Now try this

Construct the perpendicular bisector of the line $AB$.          **(2 marks)**

Remember to show **all** your construction marks.

Worked solution video

# Constructions 2

You need to know all of these constructions for your exam.

## Worked example

Use ruler and compasses to **construct** a triangle with sides of length 3 cm, 4 cm and 5.5 cm.

4 cm   3 cm

5.5 cm

**(2 marks)**

Draw one side with a ruler and label it. Then use your compasses to find the other vertex.

## Worked example

Use ruler and compasses to **construct** the bisector of angle *PQR*.

*P*

*Q*

*R*   **(2 marks)**

Mark points on each arm an equal distance from *Q*. Then use arcs to find a third point an equal distance from these two points.

## Worked example

Use ruler and compasses to **construct** a 45° angle at *P*.

45°

*P*   *M*

**(2 marks)**

Construct the perpendicular bisector of the line. Mark the midpoint *M*. Now set your compasses to the distance *PM*. Draw an arc on your bisector and join this point to *P* with a ruler.

## Worked example

Use ruler and compasses to **construct** a 60° angle at *P*.

60°

*P*   **(2 marks)**

Construct an equilateral triangle (all sides the same length). Each angle is 60°.

## Now try this

**1** Construct the bisector of the angle *ABC*.

**(2 marks)**

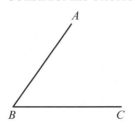

*A*

*B*   *C*

Worked solution video

**2** Use ruler and compasses to construct an angle of 30°.   **(3 marks)**

Do this on a separate piece of paper. You can construct a 30° angle by constructing a 60° angle then bisecting it.

# Loci

A **locus** is a set of points which satisfy a condition. You can construct loci using ruler and compasses. A set of points can lie inside a **region** rather than on a line or curve.

The locus of points which are 7 cm from A is the circle, centre A.

7 cm

The region of points less than 7 cm from A lies inside this circle.

The locus of points which are the **same distance** from B as from C is the perpendicular bisector of BC.

$\times^B$

$\times_C$

Points in the shaded region are closer to B than to C.

The locus of points which are 2 cm away from ST consists of two semicircles and two straight lines.

2 cm

2 cm

2 cm

The shortest distance between a point and a line is a **perpendicular** from the point to the line.

## Combining conditions

You can be asked to shade a region which satisfies more than one condition.

Here, the shaded region is more than 6 cm from point D **and** closer to line BC than to line AD.

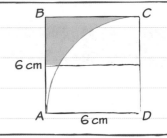

B          C

6 cm

A    6 cm    D

## Worked example

This diagram is drawn accurately and shows part of a beach and the sea.
1 cm represents 20 m.
There is a lifeguard tower at point P. Public swimming is allowed in a region of the sea less than 30 m from the lifeguard tower.
Shade this region on the diagram.            **(2 marks)**

Beach

$P_\times$

Sea

Everything in red is part of the answer.

1 cm represents 20 m so 1.5 cm represents 30 m.
There is more about scale drawing on page 99.

Set your compasses to 1.5 cm.
Set your compasses accurately by placing the point **on top** of your ruler at the 0 mark.

## Now try this

A piece of card in the shape of an equilateral triangle ABC is placed on a horizontal straight line.

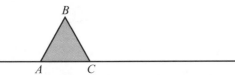

B

A    C

The card is first rotated 120° clockwise about C.
The card is then rotated 120° clockwise about B.
Draw the locus of the vertex A.            **(3 marks)**

Start with your compasses point at C and your pencil at A and draw an arc. You need to work out where B touches the ground and move your compass point to there.

# Bearings

Bearings are measured **clockwise** from **north**.

Bearings always have **three figures**. You need to add zeros if the angle is less than 100°. For instance, in this diagram the bearing of B from A is 048°.

You can measure a bearing bigger than 180° by measuring this angle and subtracting it from 360°.

The bearing of C from A is 360° − 109° = 251°

You can work out a reverse bearing by adding or subtracting 180°.

The bearing of A from B is 180° + 048° = 228°

These are alternate angles.

## Worked example

Three paths meet at O. B is due east of O.

(a) Jake walks from O to A.
What bearing does he walk on? **(1 mark)**

90° + 100° + 125° = 315°

(b) Delvinder walks from C to O.
What bearing does she walk on? **(2 marks)**

Bearing of C from O = 90° + 100°
         = 190°
Bearing of O from C = 190° − 180° = 010°

### Compass points

You need to know the compass points:

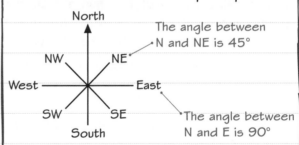

The angle between N and NE is 45°

The angle between N and E is 90°

## Problem solved!

You can work out a **reverse** bearing by adding or subtracting 180°. If the angle is greater than 180° then subtract. Remember to write your final answer as a **three-figure bearing**. You need to write 010°, not just 10°.

You will need to use problem-solving skills throughout your exam – **be prepared!**

## Now try this

An aircraft flies from A to B. The grid shows the positions of A and B.
(a) Use the grid to work out the actual distance AB.
       **(1 mark)**
(b) Measure and write down the three-figure bearing of B from A. **(1 mark)**
(c) The aircraft then flies to C. The bearing of C from A is 120°. The bearing of C from B is 080°.
Mark the position of C on the grid. **(3 marks)**

Scale: 1 cm = 100 km

# Circles

Make sure you know the definitions of **radius, diameter** and **circumference**.

Here are two different formulae for the circumference. You can use either.

Diameter = 2 × radius

**1** Circumference = π × diameter    $C = \pi d$

**2** Circumference = 2 × π × radius    $C = 2\pi r$

**LEARN IT!**

---

**π**

This symbol is the Greek letter 'pi'. It always stands for the same number.

π = 3.1415926...

Your calculator probably has a button for entering π into a calculation. You might need to press the SHIFT key first.

If your calculator leaves π in the answer, press the S⇔D button to get your answer as a decimal.

You can also use 3.142 as the value of π in your exam.

**Worked example**

Work out the circumference of this circle. Give your answer to 2 decimal places.  **(2 marks)**

6 cm

Circumference = $2\pi r$
= 2 × 3.142 × 6
= 37.704
= 37.70 cm (2 d.p.)

---

**Worked example**

2 cm    2 cm

The diagram shows an earring made from a quarter of a circle.
Work out the perimeter of the earring. Give your answer to 2 decimal places.

**(3 marks)**

Circumference of whole circle = $2\pi r$
= 2 × π × 2
= 12.5663...

Curved section of earring = 12.5663... ÷ 4
= 3.1415...

Total perimeter = 2 + 2 + 3.1415...
= 7.1415...
= 7.14 cm

Don't round your answers until the end of your calculation.

---

**Now try this**

This steering wheel has a circumference of 120 cm.
(a) Work out the diameter of the steering wheel. Give your answer to 1 decimal place.    **(2 marks)**
(b) Work out the radius of the steering wheel. Give your answer to 1 decimal place.    **(1 mark)**

**Worked solution video**

# Area of a circle

You need to know the formula for the area of a circle:

Area = $\pi \times$ radius$^2$

$A = \pi \times r \times r = \pi r^2$

**LEARN IT!**

You always need to use the **radius** when you are calculating the area.

If you are given the diameter, divide it by 2 to get the radius.

---

## Worked example

4.8 cm

Work out the area of this circle. **(2 marks)**

$$\text{Area} = \pi r^2$$
$$= \pi \times 4.8^2$$
$$= \pi \times 23.04$$
$$= 72.3822...$$
$$= 72.4 \text{ cm}^2 \text{ (1 d.p.)}$$

### Problem solved!

When you have to keep track of a lot of working, **write down** what each calculation represents. It will be really clear to you **and** the examiner what you have worked out!

> You will need to use problem-solving skills throughout your exam – **be prepared!**

## Worked example

←8 cm→

20 cm

The diagram shows two semicircles with diameters 8 cm and 20 cm.
Work out the shaded area. **(4 marks)**

Radius of large semicircle = $20 \div 2$
$\qquad\qquad\qquad\qquad\qquad = 10$

Area of large semicircle $= \frac{1}{2} \times \pi \times 10^2$
$\qquad\qquad\qquad\qquad\qquad\qquad = 157.0796...$

Radius of small semicircle $= 8 \div 2 = 4$

Area of small semicircle $= \frac{1}{2} \times \pi \times 4^2$
$\qquad\qquad\qquad\qquad\qquad\qquad = 25.1327...$

Shaded area = $157.0796... - 25.1327...$
$\qquad\qquad\quad = 131.946...$
$\qquad\qquad\quad = 132 \text{ cm}^2 \text{ (3 s.f.)}$

> The formula for the area of a circle uses the **radius**. If you are given the **diameter** you will need to divide by 2 first.

---

## Now try this

This circle and this square are equal in area.
Work out the length of the side of the square.
Give your answer correct to 1 decimal place.
**(4 marks)**

10.6 cm

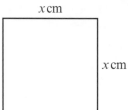

$x$ cm

$x$ cm

> Work out the area of the circle. This is the same as the area of the square. Work out the square root to find $x$.

# Sectors of circles

Each pair of radii divides a circle into two sectors, a **major sector** and a **minor sector**.

Minor sector

Major sector

You can find the area of a sector by working out what fraction it is of the whole circle.

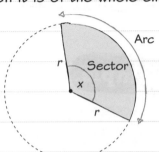

Arc

Sector

$r$

$x$

$r$

For a sector with angle $x$ of a circle with radius $r$:

Sector $= \dfrac{x}{360°}$ of the whole circle so

Area of sector $= \dfrac{x}{360°} \times \pi r^2$

Arc length $= \dfrac{x}{360°} \times 2\pi r$

**LEARN IT!**

You can give answers in terms of $\pi$.
There is more about this on the next page.

---

## Worked example

The diagram shows a minor sector of a circle of radius 13 cm.

150°

$O$

Work out the perimeter of the sector.   **(4 marks)**

Arc length $= \dfrac{x}{360°} \times 2\pi r$

$= \dfrac{150°}{360°} \times 2 \times \pi \times 13$

$= 34.03392...$

Perimeter $=$ arc length $+$ radius $+$ radius

$= 34.03392... + 13 + 13$

$= 60$ cm (2 s.f.)

Don't round until your final answer. The radius is given correct to 2 significant figures so this is a good degree of accuracy.

## Finding a missing angle

You can use the formulae for arc length or area to find a missing angle in a sector. Practise this method to help you tackle the hardest questions.

10 cm

Area = 65 cm²

$x°$

$O$

10 cm

Area of sector $= \dfrac{x}{360} \times \pi r^2$

$65 = \dfrac{x}{360} \times 100\pi$

$x = \dfrac{65 \times 360}{100\pi}$

$= 74.4845...$

$= 74.5°$ (to 3 s.f.)

---

## Now try this

$OAB$ is a sector of a circle, centre $O$.
Angle $AOB = 60°$.
$OA = OB = 12$ cm.
Work out the length of the arc $AB$.
Give your answer correct to
3 significant figures.   **(3 marks)**

You need to learn the formula for arc length.

$A$       $B$

12 cm   12 cm

60°

$O$

# Cylinders

## Surface area

To find the **surface area** of a **cylinder** you need to add up the areas of the faces.
A cylinder has two flat circular faces and one curved face. When you flatten out the curved face it is shaped like a rectangle.

Surface area = 2 × area of circle + area of rectangle

$$= 2 \times \pi r^2 + 2\pi r \times h$$
$$= 2\pi r^2 + 2\pi rh$$

**LEARN IT!**

The length of the rectangular face is the same as the circumference of the circle.

---

## Volume

For a cylinder with radius $r$ and height $h$:

Volume of cylinder = area of base × height

**LEARN IT!** = area of circle × height

$$= \pi r^2 h$$

### Problem solved!

You should always write down the formula **before** substituting. Be careful when deciding which quantities to use. You are given the **diameter** but the formula uses **radius** so you need to divide by 2.

> You will need to use problem-solving skills throughout your exam – **be prepared!**

### Worked example

This tin of soup is in the shape of a cylinder with height 11 cm. The diameter of the base is 7 cm.
$1 \text{ cm}^3 = 1 \text{ m}l$
Work out the capacity of the tin in m$l$. **(3 marks)**

Radius of base = 7 ÷ 2 = 3.5 cm
Volume = $\pi r^2 h = \pi \times 3.5^2 \times 11$
     = 423.3296... cm³
The capacity is 423 m$l$ to the nearest whole number.

---

## In terms of $\pi$

If a question asks for an **exact value** or an answer **in terms of** $\pi$ then don't use the $\pi$ button on your calculator. Write your answer as a whole number or fraction multiplied by $\pi$.

Volume of cylinder = $\pi r^2 h$
            = $\pi \times 3^2 \times 4$

**Exact answer**
Volume = $36\pi$ cm³

**Rounded answer**
Volume = 113 cm³ (to 3 s.f.)

---

### Now try this

The diagram shows an oil drum in the shape of a cylinder of height 84 cm and diameter 58 cm.
It is one-quarter full of crude oil.
Calculate the volume of oil in the cylinder.
Give your answer in litres, correct to the nearest litre.

**(4 marks)**

1 litre = 1000 cm³

# Volumes of 3D shapes

If you need to use these formulae in your exam they will be given to you with the question.

## Cone

Volume of cone
$= \frac{1}{3} \times$ area of $\times$ vertical
base height
$= \frac{1}{3}\pi r^2 h$

## Sphere

Volume of sphere $= \frac{4}{3}\pi r^3$

## Pyramid

Volume of pyramid
$= \frac{1}{3} \times$ area of $\times$ vertical
base height
$= \frac{1}{3}Ah$

---

## Worked example

The diagram shows a cone A and a cylinder B. Show that the volume of B is 8 times the volume of A. **(4 marks)**

Volume of A $= \frac{1}{3}\pi r^2 h = \frac{1}{3}\pi \times 3^2 \times 8$
$= 24\pi$

Volume of B $= \pi r^2 h = \pi \times 4^2 \times 12$
$= 192\pi$

$8 \times 24\pi = 192\pi$ so the volume of B is 8 times the volume of A.

## Problem solved!

You might have to **compare** two volumes or areas in your exam. These questions might involve
- working out the ratio between two different areas or volumes
- finding an unknown quantity represented by a letter
- finding an expression for a length, area or volume in terms of an unknown.

In this question you need to know the ratio between the two volumes. Calculate them both, then write a short **conclusion**. Make sure you show the calculation $8 \times 24\pi = 192\pi$ in your conclusion. You can leave your working in terms of $\pi$ to make it easier. There is more about this on page 107.

> You will need to use problem-solving skills throughout your exam – **be prepared!**

---

## Now try this

The radius of the base of a cone is 3 cm and its height is $h$ cm.
The radius of a sphere is 3 cm.
The volume of the cone is equal to the volume of the sphere.
Find the value of $h$. You must show your working.

**(3 marks)**

Write the expression for the volume of each shape, then set them equal to each other. Rearrange to make $h$ the subject.

# Surface area

## Cone

The formula for the **curved surface area** of a cone will be given if you need it for a question.

Curved surface area of cone = $\pi r l$

Be careful! This formula uses the slant height, $l$, of the cone.

To calculate the **total** surface area of the cone you need to add the area of the base. Surface area of cone = $\pi r^2 + \pi r l$

## Sphere

The formula for the surface area of a sphere will be given if you need to use it.

Surface area of sphere = $4\pi r^2$

For a reminder about areas of circles and surface areas of cylinders have a look at pages 105 and 107.

A hemisphere is half a sphere, so the area of the curved surface is $\frac{1}{2} \times 4\pi r^2$.

### Problem solved!

To work out the curved surface area you need to know the radius and the slant height. You are given the radius and the **vertical height**. To calculate the slant height you need to use Pythagoras' theorem. Sketch the right-angled triangle containing the missing length.

You will need to use problem-solving skills throughout your exam – **be prepared!**

### Worked example

The diagram shows a cone with vertical height 12 cm and base radius 5 cm. Work out the curved surface area of the cone.
**(4 marks)**

$l^2 = 12^2 + 5^2 = 169$
$l = 13$ cm
Curved surface area
$= \pi r l$
$= \pi \times 13 \times 5$
$= 65\pi$ cm$^2$

### Compound shapes

You can calculate the surface area of more complicated shapes by adding together the surface area of each part.

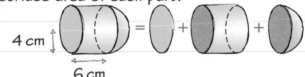

Surface area $= \pi(4)^2 + 2\pi(4)(6) + \frac{1}{2}[4\pi(4)^2]$
$= 96\pi$ cm$^2$

### Now try this

1 The diagram shows an object made from two cones, placed one on top of the other.
The top cone has a slant height of 6 cm and the bottom cone has a slant height of 7 cm.
Both cones have a radius of 4 cm.
Work out the total surface area of the object. Give your answer in terms of $\pi$. **(4 marks)**

2 A solid object is formed by joining a hemisphere to a cylinder.
Both the hemisphere and the cylinder have a radius of 2.1 cm. The cylinder has a height of 5.6 cm. Work out the total surface area of the object. Give your answer to 3 significant figures. **(4 marks)**

# Similarity and congruence

If one shape is an enlargement of another, the shapes are **similar**.

These triangles are similar.

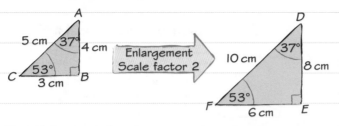

Enlargement Scale factor 2

The angles in similar shapes are the same.

**Congruent** shapes are exactly the same shape and size. They have the same area and the same perimeter.

Rotations and reflections are congruent.

Enlargements are a different size so they are not congruent.

## Worked example

Here is a large triangle made from 16 equilateral triangles.

Everything in red is part of the answer.

Shade a shape which is **similar** to the shaded shape but **not** congruent.          **(2 marks)**

Congruent shapes are exactly the same shape and size. Similar shapes can be enlargements. You can shade in a rhombus which is an enlargement of the grey shape with a scale factor of 2.

---

## Worked example

These two cubes are similar.

2 cm          8 cm

How many times will the 2 cm cube fit inside the 8 cm cube?          **(2 marks)**

4 × 4 × 4 = 64

## Problem solved!

Imagine placing copies of the small cube inside the larger cube. A sketch might help. You could fit 4 × 4 = 16 across the bottom. You can fit 4 layers like this, so in total you can fit 4 × 4 × 4 copies of the small cube inside the larger cube.

You will need to use problem-solving skills throughout your exam – **be prepared!**

---

## Now try this

Here are some shapes on a grid.
Write down three pairs of similar shapes.          **(3 marks)**

If one shape is an enlargement of another, the shapes are similar. Similar shapes are not necessarily congruent.

# Similar shapes

You need to recognise similar triangles and find missing lengths and angles in similar shapes.

**Similar triangles** satisfy one of these three conditions:

 **1** All three pairs of angles are equal.

 **2** All three pairs of sides are in the same ratio.

 **3** Two sides are in the same ratio and the included angle is equal.

---

## Worked example

*XYZ* and *ABC* are similar triangles.

(a) Work out the length of *AC*. **(2 marks)**

$$\frac{AC}{XZ} = \frac{BC}{YZ}$$

$$\frac{AC}{8} = \frac{7.8}{6}$$

$$AC = \frac{7.8 \times 8}{6}$$

$$= 10.4 \text{ cm}$$

(b) Work out the length of *XY*. **(2 marks)**

$$\frac{XY}{AB} = \frac{YZ}{BC}$$

$$\frac{XY}{6.5} = \frac{6}{7.8}$$

$$XY = \frac{6 \times 6.5}{7.8}$$

$$= 5 \text{ cm}$$

> Start with the unknown length on top of a fraction. Make sure you write your ratios in the correct order.

### Similar shapes checklist

Use these facts to solve similar shapes problems:

- ✓ Corresponding angles are equal.
- ✓ Corresponding sides are in the same ratio.

### Spotting similar triangles

Here are some similar triangles:

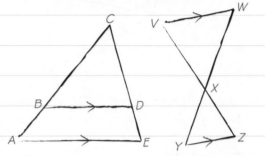

Triangle *ACE* is similar to triangle *BCD*.

Triangle *VWX* is similar to triangle *ZYX*.

---

## Now try this

Triangles *ABC* and *PQR* are similar.
Angle *ACB* = angle *PRQ*.

(a) Work out the size of angle *PRQ*. **(2 marks)**

(b) Work out the length of *PQ*. **(2 marks)**

# Congruent triangles

Triangles are **congruent** if **any one** of these four conditions is true.

**1** **SSS** (three sides are equal)

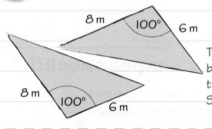

2.4 cm
3 cm
1.8 cm

1.8 cm
3 cm
2.4 cm

**2** **AAS** (two angles and a corresponding side are equal)

95°  40°
3 cm

40°
3 cm  95°

**3** **SAS** (two sides and the included angle are equal)

8 m  100°  6 m

8 m  100°  6 m

The angle must be **between** the two sides for SAS.

**4** **RHS** (right angle, hypotenuse and a side are equal)

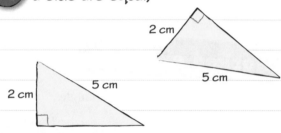

2 cm
5 cm

2 cm  5 cm

## Worked example

A
6 cm
D
B
6 cm
C

Triangle *ABD* and triangle *BDC* are congruent. Explain which information in the diagram supports this statement. **(3 marks)**

*ABD* and *BDC* are both right-angled triangles. The hypotenuse of both triangles is 6 cm. Side *BD* is common to both triangles so is equal. Therefore the triangles satisfy the RHS condition and are congruent.

## Problem solved!

To explain why the two triangles are congruent, you need to show that **one** of the conditions above is satisfied. Once you have decided which one, explain which lengths or angles are the same and why. Then write down **which condition** is satisfied. It's OK to use the abbreviations above.

You will need to use problem-solving skills throughout your exam – **be prepared!**

### Common sides

If two triangles have a side in **common** then those two sides are equal.

A
P  C
B

## Now try this

In the quadrilateral *ABCD*, *BC = CD* and the angles at *B* and *D* are right angles.

Show that triangle *ABC* is congruent to triangle *ADC*. **(3 marks)**

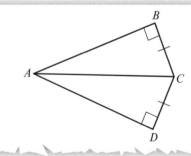

B
A  C
D

# Vectors

A vector has a **magnitude** (or size) and a **direction**.

This vector can be written as a, $\overrightarrow{AB}$ or $\begin{pmatrix} 2 \\ 5 \end{pmatrix}$

You can multiply a vector by a number. The new vector has a different length but the same direction.

If **b** is a vector then −**b** is a vector with the same length but opposite direction.

---

## Worked example

In the diagram, *OADB* and *ACED* are two identical parallelograms.

**a** is the vector $\overrightarrow{OA}$
**b** is the vector $\overrightarrow{OB}$

Find the following vectors in terms of **a** and **b**.

(a) $\overrightarrow{OD}$           **(1 mark)**

$\overrightarrow{OD} = \overrightarrow{OA} + \overrightarrow{AD}$
$= a + b$

(b) $\overrightarrow{EB}$           **(1 mark)**

$\overrightarrow{EB} = \overrightarrow{ED} + \overrightarrow{DB}$
$= -a + -a = -2a$

(c) $\overrightarrow{BC}$           **(1 mark)**

$\overrightarrow{BC} = \overrightarrow{BD} + \overrightarrow{DE} + \overrightarrow{EC}$
$= a + a - b = 2a - b$

## Adding vectors

You can add vectors using the **triangle law**. You trace a path along the added vectors to find the new vector.

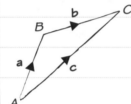

$a + b = c$

**c** is the resultant vector of **a** and **b**.

If $a = \begin{pmatrix} 2 \\ 4 \end{pmatrix}$ and $b = \begin{pmatrix} 6 \\ 3 \end{pmatrix}$

then $c = \begin{pmatrix} 2 + 6 \\ 4 + 3 \end{pmatrix} = \begin{pmatrix} 8 \\ 7 \end{pmatrix}$

> For each vector, trace a path along the shape from the start point to the end point. If you go in the **opposite** direction to the vector then you need to **subtract**. $\overrightarrow{AD} = \overrightarrow{OB}$ because they are **parallel**.

> Simplify your vectors as much as possible.

---

If you multiply a column vector by a number you have to multiply **both** parts.

$2 \times \begin{pmatrix} p \\ q \end{pmatrix} = \begin{pmatrix} 2p \\ 2q \end{pmatrix}$

To **add** column vectors you add the top numbers and add the bottom numbers:

$\begin{pmatrix} c \\ d \end{pmatrix} + \begin{pmatrix} e \\ f \end{pmatrix} = \begin{pmatrix} c + e \\ d + f \end{pmatrix}$

## Now try this

The vectors **a** and **b** are defined as

$a = \begin{pmatrix} 3 \\ 5 \end{pmatrix}$      $b = \begin{pmatrix} 2 \\ -9 \end{pmatrix}$

Write the following as column vectors.

(a) 2**a**                  **(1 mark)**
(b) **a** + **b**            **(1 mark)**
(c) **b** − 3**a**          **(3 marks)**

# Problem-solving practice 1

About half of the questions in your Foundation GCSE exam will require you to **problem-solve, reason, interpret** or **communicate** mathematically. If you come across a tricky or unfamiliar question in your exam you can try some of these strategies:

☑ Sketch a diagram to see what is going on.

☑ Try the problem with smaller or easier numbers.

☑ Plan your strategy before you start.

☑ Write down any formulae you might be able to use.

☑ Use *x* or *n* to represent an unknown value.

---

**1** This diagram shows a quadrilateral.

Work out the value of *y*.          (3 marks)

### Angles 2 page 75
### Using algebra page 52

Start by writing an expression for the sum of the angles in the quadrilateral. Use angle facts to write down what this sum is equal to. This gives you an equation you can solve to find the value of *y*.

**TOP TIP**

If you know what unknown angles should add up to, then you can write an equation and solve it to find the value of the unknown.

---

**2** A buoy is 6 km from a ship on a bearing of 290°.

A lighthouse is 8 km east of the ship.

Work out the distance between the buoy and the lighthouse.          (3 marks)

### Bearings page 103
### Scale drawings and maps page 99

You'll need to draw a scale diagram to solve this question. A good scale to use would be 1 cm = 2 km. Always use a ruler and a sharp pencil to draw any lines and **don't** rub out any construction lines or working.

**TOP TIP**

Bearings less than 180° are to the **right** of North. Bearings between 180° and 360° are to the **left**. Bearings always have three figures.

# Problem-solving practice 2

**3** The diagram shows a semicircle. It has two identical semicircles drawn inside it.

Show that the shaded area is exactly half of the area of the large semicircle. **(3 marks)**

## Area of a circle page 105

The easiest way to tackle this problem is to choose a value for the diameter of each semicircle. The two small semicircles are identical, so if they both have diameter 2 cm, the large semicircle will have diameter 4 cm. Work out the area of each semicircle using these values, then write a conclusion. Make sure all your working is clear and neat.

**TOP TIP**

If you aren't given any lengths on a diagram you might be able to:
- use $x$ to represent one of the lengths and write the other lengths in terms of $x$
- choose some easy numbers yourself for the lengths.

---

**4** A ladder is 6 m long.

The ladder is placed on horizontal ground, resting against a vertical wall.

The instructions for using the ladder say that the bottom of the ladder must not be closer than 1.5 m to the bottom of the wall.

How far up the wall can the ladder reach if the instructions are followed? **(3 marks)**

## Pythagoras' theorem page 91

You should definitely draw a sketch to show the information in the question.

**TOP TIP**

Be careful when you are working out the length of a **short** side using Pythagoras' theorem.

Remember: $\text{short}^2 + \text{short}^2 = \text{long}^2$

$\text{short}^2 = \text{long}^2 - \text{short}^2$

---

**5**

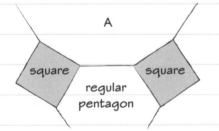

The diagram shows 2 squares, part of a regular pentagon and part of a regular $n$-sided polygon, A.

Calculate the value of $n$.

Show your working clearly. **(5 marks)**

## Angles in polygons page 77

Follow these steps:
1. Work out the interior angles of a regular pentagon and a square.
2. Use the fact that the angles around a point add up to 360° to find the interior angle of A.
3. Subtract this from 180° to find the exterior angle of A.
4. Divide 360° by this to work out $n$.

**TOP TIP**

If there are a lot of steps in a question it's a good idea to **plan** your answer before you start.

# Two-way tables

You can answer questions about two-way tables by adding or subtracting.

|  | Year 7 | Year 8 | Year 9 | Total |
|---|---|---|---|---|
| Vegetarian | 14 | 22 | 25 | 61 |
| Not vegetarian | 72 | 63 | 54 | 189 |
| Total | 86 | 85 | 79 | 250 |

There were 61 vegetarians in total.

In total 250 students were surveyed.

There were 86 Year 7 students surveyed.

There were 63 non-vegetarians in Year 8.

## Worked example

Anton surveyed 120 people about how they voted at the last general election. He recorded the results in a two-way table:

|  | Labour | Conservative | Other | Total |
|---|---|---|---|---|
| Female | 21 | 13 | 13 | 47 |
| Male | 32 | 27 | 14 | 73 |
| Total | 53 | 40 | 27 | 120 |

Complete the two-way table. **(4 marks)**

Labour column: 53 − 21 = 32

Female row: 47 − 21 − 13 = 13

Conservative column: 13 + 27 = 40

Total row: 120 − 53 − 40 = 27

Other column: 27 − 13 = 14

Male row: 32 + 27 + 14 = 73

Check: 47 + 73 = 120

53 + 40 + 27 = 120 ✓

Everything in red is part of the answer.

### Golden rules

The numbers in each column add up to the total for that column.

| Other |
|---|
| 13 |
| + 14 |
| = 27 |

The numbers in each row add up to the total for that row.

| Female | 21 | + 13 | + 13 | = 47 |
|---|---|---|---|---|

To complete a two-way table:
- write the total in the bottom right-hand cell
- look for rows and columns with only one missing number
- use addition and subtraction to find any missing values
- fill in the missing values as you go along.

### Check it!
Add up the row totals and the column totals. They should be the same.

## Now try this

A photographic shop offers prints in three different sizes and on three different types of paper.

This two-way table shows information about the choices made by customers on one day.

Complete the two-way table. **(4 marks)**

**Worked solution video**

|  | Glass | Matt | Lustre | Total |
|---|---|---|---|---|
| Small | 20 | 35 | 12 |  |
| Medium | 63 |  | 29 |  |
| Large |  | 24 |  |  |
| Total | 105 |  | 59 | 325 |

Look for rows or columns with only one missing number.

# Pictograms

A **pictogram** can be used to represent data from a tally chart or frequency table.

This pictogram shows the results of a survey about how people watch television.

There is one row for each option.

Key: 📺 represents 2 people

A pictogram must have a **key**. This tells you how many items are represented by each picture.

| Freeview | 📺 📺 📺 📺 📺 📺 |
| Satellite | 📺 📺 📺 📺 📺 |
| Cable | 📺 |
| Internet | 📺 📺 📺 📺 |

📺 represents 2 people so 12 people said they watched television using Freeview.

Each television represents 2 people, so half a television represents 1 person. This row represents 9 people.

To work out the total number of people in the survey, add together the totals of each row: $12 + 9 + 2 + 7 = 30$.

## Worked example

The pictogram shows the numbers of films Adam watched in June, July and August.

| June | ⊞ ⊞ |
| July | ⊞ ⊟ |
| August | ⊞ □ |
| September | ⊞ ⊞ ⊟ |

Everything in red is part of the answer.

Key: ⊞ represents 4 films

(a) Write down the number of films Adam watched in June. **(1 mark)**

8

(b) Write down the number of films Adam watched in August. **(1 mark)**

5

In September Adam watched 10 films.

(c) Use this information to complete the pictogram. **(1 mark)**

Use the key to work out what each picture represents.

⊞ = 4 films    ⊟ = 3 films
⊟ = 2 films    □ = 1 film

There is a block of 4 squares and a block of 1 square in August. This represents $4 + 1 = 5$ films.

To represent 10 films you need two blocks of 4 squares and one block of 2 squares. Draw the squares neatly on the pictogram.

## Tallies

A **tally** is a bit like a pictogram:

I represents 1        ⅢⅠ represents 5

So to represent 12 you would draw:

ⅢⅠ ⅢⅠ II

## Now try this

Worked solution video

A magazine uses a pictogram to show the average house price in five different English counties.
The key for the pictogram is missing.
The average house price in Herefordshire is £225 000.
The average house price in Lincolnshire is £125 000.

(a) Complete the pictogram to show this information. Include a key for the pictogram. **(2 marks)**

(b) Work out the average house price in Essex. **(1 mark)**

Key

| Lancashire | 🏠 🏠 🏠 |
| Herefordshire | 🏠 🏠 🏠 🏠 🏠 |
| Essex | 🏠 🏠 🏠 🏠 🏠 |
| Surrey | 🏠 🏠 🏠 🏠 🏠 🏠 🏠 🏠 🏠 |
| Lincolnshire | |

# Bar charts

You can use a **bar chart** to represent data given in a tally chart or frequency table. This **dual bar chart** shows the numbers of pairs of jeans owned by the members of a class.

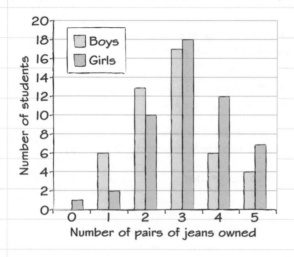

## Bar chart features

- ✓ Bars are the same width.
- ✓ There is a gap between the bars.
- ✓ Both axes have labels.
- ✓ Bars can be drawn horizontally or vertically.
- ✓ The height (or length) of each bar represents the frequency.
- ✓ In a dual bar chart two (or more) bars are drawn side by side. They can be used to compare data.

## Worked example

Kaitlyn carried out a survey of the colours of cars which passed the school gate in 10 minutes. Here are her results.

| Colour | Tally |
|--------|-------|
| Red | 𝍸𝍸𝍸 \|\|\|\| |
| Black | 𝍸𝍸𝍸 \|\| |
| Silver | 𝍸𝍸𝍸 𝍸𝍸𝍸 |
| Blue | \|\|\| |

Use the grid to draw a <u>suitable chart or diagram</u> to represent Kaitlyn's results.    **(4 marks)**

The question says 'suitable chart or diagram'. You could also represent this data using a pictogram or pie chart.

## Now try this

Worked solution video

The bar chart shows the number of job applications Susan sent during one week.

On Tuesday Susan sent 12 job applications.

How many job applications did Susan send altogether during this week?
You must show your working.    **(3 marks)**

Start by using the information in the question to work out the scale on the vertical axis.

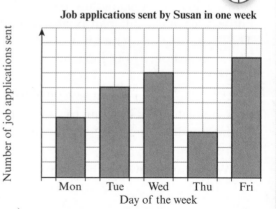

# Pie charts

In your exam you might have to interpret information given in a **pie chart**, or draw a pie chart from a frequency table. This pie chart shows the favourite sports of 120 students.

There are 360° in a circle and 120 students. 360 ÷ 120 = 3

This means that each student is represented by an angle of 3°.

66 ÷ 3 = 22
So 66° represents 22 students.

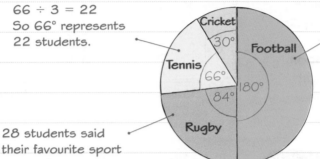

Half the students chose football. This **sector** of the pie chart represents 60 students.

28 students said their favourite sport was rugby.

---

## Worked example

You can revise measuring and drawing angles using a protractor on page 96.

A farm has 40 fruit trees.
The table shows the number of each type of tree.
Draw a pie chart to represent this information.

| Type of fruit tree | Number of trees | Angle |
|---|---|---|
| Apple | 12 | 12 × 9° = 108° |
| Plum | 5 | 5 × 9° = 45° |
| Pear | 14 | 14 × 9° = 126° |
| Peach | 9 | 9 × 9° = 81° |

Angle for 1 tree = 360° ÷ 40 = 9°
Check: 108° + 45° + 126° + 81° = 360° ✓

Everything in red is part of the answer.

**(4 marks)**

You need a sharp pencil, compasses and a protractor to draw a pie chart.
1. Add an 'Angle' column to the frequency table.
2. There are 360° in a full circle. There are 40 trees. So divide 360° by 40 to find the angle that represents 1 tree.
3. Multiply the angle that represents 1 tree by the number of each tree type to find the angle for each type.
4. **Check** that your angles add up to 360°.
5. Draw a circle using compasses. Draw a vertical line from the centre to the edge of the circle. Use a protractor to measure and draw the first angle (108°) from this line. Draw each angle carefully in order.
6. Label each sector of your pie chart with the type of fruit tree.

---

## Now try this

A school canteen offered four lunch options on one day. This table shows the choices made by 90 students.

Draw a pie chart to represent this data.

**(4 marks)**

Start by drawing a circle with a radius of at least 3 cm.

Worked solution video

| Lunch choice | Number of students |
|---|---|
| Salad | 21 |
| Stew | 9 |
| Pizza | 45 |
| Curry | 15 |

# Scatter graphs

The points on a scatter graph aren't always scattered. If the points are almost on a straight line then the scatter graph shows **correlation**. The better the straight line, the stronger the correlation.

Negative correlation

No correlation

Positive correlation

An **outlier** is a value that does not fit the pattern of the data.

## Worked example

This scatter graph shows the engine capacity of some cars and the distance they will travel on one gallon of petrol.

(a) Estimate how far a car with a 3-litre engine will travel on 1 gallon of petrol. **(2 marks)**

17 miles

(b) Comment on the reliability of your estimate. **(1 mark)**

3 litres is within the range of the data values (interpolation) and the correlation is strong, so the estimate is reliable.

## Cause and effect

Watch out! Correlation doesn't always mean that two variables are related.

Bottled water doesn't cause bee stings but, when the weather is hotter, bottled water sales and bee stings both increase.

Use a **line of best fit** to estimate. Draw a straight line as close to as many of the points on the graph as possible.

## Estimating and predicting

✓ If you are predicting a value that is **within** the range of the data your prediction will be **more accurate**.
This is called **interpolation**.

✗ If you are predicting a value that falls **outside** the range of the data your prediction will be **less accurate**.
This is called **extrapolation**.

## Now try this

This scatter graph shows the daily hours of sunshine and the daily maximum temperature at 10 seaside resorts in England one day last summer.

(a) Another resort had 10 hours of sunshine each day. Use a line of best fit to predict the maximum temperature at this resort. **(2 marks)**

(b) Comment on the reliability of your estimate. **(1 mark)**

# Averages and range

There are three different types of average: the mean, median and mode.
The range of a set of data tells you how spread out the data is.

## Worked example

The **mode** is the value which occurs **most often**.

Here are six numbers.
4  5  9  7  4  4

(a) Write down the mode. **(1 mark)**
The mode is 4.

(b) Work out the mean. **(2 marks)**
4 + 5 + 9 + 7 + 4 + 4 = 33
33 ÷ 6 = 5.5
The mean is 5.5

To find the **mean** you add together all the numbers and then divide by how many numbers there are. Don't round your answer.

The **median** is the middle value. First write the values in order from smallest to largest. If there are two middle values, the median is halfway between them.

(c) Work out the median. **(2 marks)**
4   4  ④  ⑤  7   9
The median is 4.5

(d) Work out the range. **(1 mark)**
9 – 4 = 5
The range is 5.

Range = largest value – smallest value

## Worked example

Kayla has three numbered cards.
The numbers are hidden.

The mode of the three numbers is 5.
The mean of the three numbers is 4.
Work out the three numbers on the cards.
**(3 marks)**

Mode = 5
At least two of the cards are 5s.
Mean = 4
Sum of three cards = 4 × 3 = 12
5 + 5 + ? = 12
The other card is a 2.
The cards are 5, 5 and 2.

## Problem solved!

The mode is the most common value. There are three cards so at least two must have 5 written on them.

To find the other value you can use this formula:
**Sum of values = mean × number of values**

**Check it!**
Calculate the mean of your three values:
5 + 5 + 2 = 12
12 ÷ 3 = 4 ✓

> You will need to use problem-solving skills throughout your exam – **be prepared!**

## Now try this

Mrs Miller gives 11 students in her class a mental arithmetic test.
The test is marked out of 15. These are their marks:

   12   13   8   14   11   8   13   14   15   14   10

Which **one** of these statements is **not** true? **(1 mark)**

The median is 13 marks ☐    The mean is larger than the mode ☐
The range is half the mode ☐    The mode is larger than the median ☐

Work out all three averages and the range before answering the question.

# Averages from tables 1

You need to be really careful when you are calculating averages from a frequency table. Remember that the **frequency** is not a data value – it tells you the **number** of pieces of data with a given value.

## Worked example

Leah asked 40 people how many pets they owned. The table shows her results.

| Number of pets $x$ | Frequency $f$ | Frequency × number of pets $f \times x$ |
|---|---|---|
| 0 | 13 | $13 \times 0 = 0$ |
| 1 | 18 | $18 \times 1 = 18$ |
| 2 | 7 | $7 \times 2 = 14$ |
| 3 | 2 | $2 \times 3 = 6$ |

(a) Write down the mode.                  (1 mark)

The mode is 1 pet.

(b) Write down the range.                 (1 mark)

The range is 3 − 0 = 3 pets.

(c) Work out the median.                  (1 mark)

The median is 1 pet.

(d) Work out the mean.                    (3 marks)

Total of $f \times x$ column = 0 + 18 + 14 + 6 = 38
Total frequency = 13 + 18 + 7 + 2 = 40
38 ÷ 40 = 0.95
The mean is 0.95 pets.

> Everything in red is part of the answer.

(a) The **mode** is the value with the highest frequency. The highest frequency is 18.

(b) The **range** is the difference between the highest and lowest values. So 3 − 0 = 3 pets.

(c) The **median** is the middle value. There are 40 values so the median is halfway between the 20th and 21st values. The first 13 values are all 0 and the next 18 values are all 1. This means that the 20th and 21st values are both 1, so the median is 1 pet.

(d) To calculate the **mean** from a frequency table you need to add an extra column. Label your new column 'Frequency × number of pets' or '$f \times x$'. The total in the $f \times x$ column represents the total number of pets owned (38 pets).
Use this rule to work out the mean:

$$\text{Mean} = \frac{\text{total of } (f \times x) \text{ column}}{\text{total frequency}}$$

Do not round your answer.

---

## Now try this

The table shows the number of goals scored by 50 teams in matches taking place one weekend.
(a) Write down the modal number of goals scored.                  (1 mark)
(b) Work out the median number of goals scored.                  (1 mark)
(c) Work out the mean number of goals scored.                  (3 marks)

| Goals scored ($x$) | Frequency ($f$) | |
|---|---|---|
| 0 | 7 | |
| 1 | 15 | |
| 2 | 14 | |
| 3 | 7 | |
| 4 | 4 | |
| 5 | 3 | |

Use the extra column for 'Frequency × Goals scored ($f \times x$)'.

# Averages from tables 2

Sometimes data in a frequency table is grouped into **class intervals**. You don't know the exact data values, but you can calculate an **estimate** of the mean, and write down which class interval contains the median and which one has the highest frequency.

## Worked example

Maisie recorded the times, in minutes, taken by 150 students to travel to school. The table shows her results.

| Time (*t* minutes) | Frequency *f* | Midpoint *x* | *f* × *x* |
|---|---|---|---|
| $0 \leqslant t < 20$ | 65 | 10 | 65 × 10 = 650 |
| $20 \leqslant t < 40$ | 42 | 30 | 42 × 30 = 1260 |
| $40 \leqslant t < 60$ | 39 | 50 | 39 × 50 = 1950 |
| $60 \leqslant t < 80$ | 4 | 70 | 4 × 70 = 280 |

Everything in red is part of the answer.

(a) Write down the modal class interval. **(1 mark)**
$0 \leqslant t < 20$

(b) Write down the class interval which contains the median. **(1 mark)**
$20 \leqslant t < 40$

(c) Work out an estimate for the mean number of minutes that the students took to travel to school. **(4 marks)**

Sum of *f* × *x* column
= 650 + 1260 + 1950 + 280 = 4140

Total frequency
= 65 + 42 + 39 + 4 = 150

4140 ÷ 150 = 27.6

Estimated mean = 27.6 minutes.

1. Add extra columns to the table.
2. Use the first extra column for 'midpoint' and work out the midpoint of each class interval.
3. Label 'frequency' *f* and 'midpoint' *x*.
4. Use the final column for *f* × *x*. Make sure you use the midpoint when calculating *f* × *x* for each row.
5. Use this rule to estimate the mean.

$$\frac{\text{Estimate}}{\text{of mean}} = \frac{\text{Total of } (f \times x \text{ column})}{\text{Total frequency}}$$

(d) Explain why your answer to part (c) is an estimate. **(1 mark)**
Because you don't know the exact data values.

---

## Now try this

The table shows the marks obtained by 50 students in a maths test.
(a) Write down the modal class interval. **(1 mark)**
(b) Write down the class interval that contains the median. **(1 mark)**
(c) Work out an estimate of the mean mark for these students. **(4 marks)**

| Marks (*x*) | Frequency (*f*) |
|---|---|
| $0 < x \leqslant 10$ | 10 |
| $10 < x \leqslant 20$ | 11 |
| $20 < x \leqslant 30$ | 8 |
| $30 < x \leqslant 40$ | 6 |
| $40 < x \leqslant 50$ | 10 |
| $50 < x \leqslant 60$ | 5 |

# Line graphs

A **time series** graph is a line graph that shows how a variable changes over a period of time. This time series graph shows median incomes in the UK from 1980 to 2010.

The graph should have a title. •————→ UK median incomes

The **vertical axis** shows the variable which changes over time.

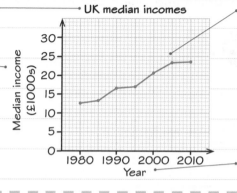

The graph shows an **upward trend**. This means that in general, as time passes, the median income **increases**.

The **horizontal axis** always shows units of time.

---

## Vertical line graphs

A vertical line graph is sometimes called a **bar-line graph**. It works in the same way as a **bar chart** but the bars are straight, vertical lines.

You can revise bar charts on page 118.

## Problem solved!

This is like finding the mean from this frequency table:

| Number of pairs of trainers | 0 | 1 | 2 | 3 | 4 | 5 | 6 |
|---|---|---|---|---|---|---|---|
| Frequency | 0 | 4 | 7 | 6 | 2 | 0 | 1 |

Work out the total number of pairs of trainers owned, then divide by the total number of members of the class.

You can revise how to find the mean from a frequency table on page 122.

> You will need to use problem-solving skills throughout your exam – **be prepared!**

## Worked example

This vertical line graph shows the number of pairs of trainers owned by the members of a class.

Number of pairs of trainers

What was the mean number of pairs of trainers owned? **(2 marks)**

Total number of pairs of trainers owned
$0 \times 0 + 4 \times 1 + 7 \times 2 + 6 \times 3 +$
$\qquad 2 \times 4 + 0 \times 5 + 1 \times 6 = 50$

Total number of members of class
$0 + 4 + 7 + 6 + 2 + 0 + 1 = 20$
$50 \div 20 = 2.5$

The mean is 2.5 pairs of trainers.

---

## Now try this

The table on the right gives information about the percentage of students who passed a professional exam each year from 2007 to 2013.

| Year | 2007 | 2008 | 2009 | 2010 | 2011 | 2012 | 2013 |
|---|---|---|---|---|---|---|---|
| Percentage (%) | 58 | 60 | 53 | 55 | 63 | 65 | 73 |

(a) Draw a time series graph to represent this data. **(3 marks)**

(b) Describe the trend. **(1 mark)**

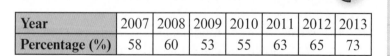

You can use a 'break' in your vertical axis.

# Sampling

In statistics, a **population** is a group of people you are interested in. A **sample** is a smaller group chosen from a larger population. You can use data from the sample to make **predictions** about the whole population.

## Advantages of sampling

✓ It is cheaper to survey a sample than a whole population.

✓ It is quicker to collect data from a sample.

✓ It is easier to analyse data from a sample and calculate statistics.

## Random sample

In a random sample, every member of the population has an equal chance of being included in the sample. Here are two ways of selecting a random sample:

**1** Put the names of every member of the population in a hat and select your sample at random.

**2** Assign a number to every member of the population and choose random numbers using a computer or calculator.

## Problem solved!

If a sample is too small the results can be **biased**. This means that the sample is not representative of the whole population, and predictions about the population won't be very accurate. There are two reasons why Ashik's sample might be biased:

| Problem | Solution |
|---------|----------|
| Ashik only selected a sample of 5 students to represent his entire school. | Choose a **larger** sample. Ashik should have surveyed at least 20 or 30 students. |
| All the students in Ashik's sample were in one class, so were in the same year group. | Choose a **random sample**. If everyone has an equal chance of being selected there will be a range of ages. |

You will need to use problem-solving skills throughout your exam – **be prepared!**

## Worked example

Ashik wants to investigate the number of hours of television students in his school watch each week. He surveys five members of his class and gets these results:

15    6    22    11    18

(a) Use this data to estimate the mean number of hours of television watched each week by students in Ashik's school. **(2 marks)**

15 + 6 + 22 + 11 + 18 = 72
72 ÷ 5 = 14.4
An estimate for the mean of the population is 14.4 hours.

(b) Comment on the reliability of your estimate. **(1 mark)**

The estimate is not very reliable because of the small sample size.

(c) How could Ashik reduce bias in his sample? **(2 marks)**

Select a larger sample, and select a random sample from the whole population (school).

## Now try this

Amy and Paul are investigating whether trains at their local station run on time. This table summarises their results.

| | Amy | Paul |
|---|---|---|
| Number of trains observed | 6 | 25 |
| Number of trains late | 3 | 8 |

Amy says 'There is a 50:50 chance that a train will be late.'

(a) Comment on the reliability of Amy's statement. **(1 mark)**

(b) Use Paul's results to estimate the probability of a train being late. **(1 mark)**

There is more about probability on page 130.

# Comparing data

You can use averages like the **mean** or **median** and measures of spread like the **range** to compare two sets of data. Follow these steps:

**1** Calculate the same average and the range for both data sets.

**2** Write a sentence for each statistic **comparing** the values for each data set.

**3** Only make a statement if you can back it up with **statistical evidence**.

## Using graphs and charts

Some graphs and charts are particularly useful for comparing data.

☑ A **dual bar chart** lets you compare two frequency distributions.

☑ Two **pie charts** can be used to compare **proportions**.

☑ You can plot two **time series** graphs on the same axes to compare how two different sets of data vary over time.

## Worked example

Melissa and Fran want to compare their long jump distances. They each jump five times. Fran's distances are:

263 cm   194 cm   220 cm   305 cm   280 cm

Melissa's distances have a mean of 292 cm and a range of 185 cm.

Compare the distances for Melissa and Fran.

**(4 marks)**

263 + 194 + 220 + 305 + 280 = 1262

Fran's mean = 1262 ÷ 5 = 252.4 cm

Fran's range = 305 − 194 = 111 cm

Melissa jumped further on average because she had a larger mean.

Fran's jumps were more consistent because she had a smaller range.

You need to interpret your results in the context of the question. Write one sentence for the mean and one sentence for the range.

You are given the mean and the range for Melissa's distances, so you need to calculate these statistics for Fran's distances.
There is more about calculating the mean and the range on page 121.

## Writing conclusions

Here are some examples of good sentences comparing data:

> Class A had more consistent exam results because they had a smaller range.

> Trees in Park B are shorter on average than trees in Park A (smaller median).

> The medians were similar, so on average the apples from both farms were the same weight.

Remember statistics can be the same as well as different.

## Now try this

Anna and Carla are in the same maths class. Every week they take a 20-mark test.
Here are Anna's first eight test scores.

11   15   9   12   13   16   13   15

Carla's mean score in these tests is 11.

Carla's highest score was 14 and her lowest score was 10.

Compare Anna and Carla's test scores.   **(4 marks)**

Start by working out Anna's mean score, and the ranges of both sets of scores.

# Probability 1

The probability that an event will happen is a value from 0 to 1.

The probability tells you how likely the event is to happen.

An event that is **certain** to happen has a probability of 1.

An event that is **impossible** has a probability of 0.

You can write a probability as a fraction, a decimal or a percentage.

Impossible     Even chance     Certain

| Fraction | Decimal | Percentage |
|----------|---------|------------|
| $\frac{1}{2}$ | 0.5 | 50% |

## Worked example

It is **very likely** that it will rain in Newcastle next October. Put a cross near 1 on the probability scale.

(a) On this probability scale, mark with a × the probability that it will rain in Newcastle next October. **(1 mark)**

0                     1

(b) Isobel says the probability she will be late for school is 7. Explain why Isobel is wrong. **(1 mark)**

Probabilities are numbers from 0 to 1.

### Writing probabilities

The probability of rolling a 6 is $\frac{1}{6}$

You can write $P(6) = \frac{1}{6}$

There is one 6. There are six possible outcomes: 1, 2, 3, 4, 5, 6.

The probability of a coin landing heads up is $\frac{1}{2}$. You can write $P(\text{Heads}) = \frac{1}{2}$

There is one head. There are two possible outcomes: heads or tails.

## Worked example

This spinner has eight equal sections.

The spinner is spun.

(a) Write down the probability that the spinner will land on the letter Z. **(1 mark)**

$\frac{3}{8}$

(b) Use a suitable probability **word** to complete this sentence:

The chance of the spinner landing on X is

evens. **(1 mark)**

### Golden rule

$$\text{Probability} = \frac{\text{number of successful outcomes}}{\text{total number of possible outcomes}}$$

Half the sections on the spinner have the letter X, so there is an even chance of landing on X. Here are some probability words you can use in your exam:

Likely   Unlikely   Evens   Certain   Impossible

## Now try this

Count the number of successful outcomes in each case.

A letter is chosen at random from this word.

| P | H | Y | S | I | C | I | S | T | S |
|---|---|---|---|---|---|---|---|---|---|

Work out the probability that the letter will be

(a) I **(1 mark)**  (b) S or Y or C **(1 mark)**  (c) M. **(1 mark)**

Worked solution video

# Probability 2

The probabilities (P) of all the different outcomes of an event add up to 1.

If you know the probability that something will happen, you can calculate the probability that it won't happen.

P(event doesn't happen) = 1 − P(event does happen)

This spinner will definitely land on either red, yellow or green. So the probability of this happening is 1.

The probability of rolling a 6 on a normal fair dice is $\frac{1}{6}$. So the probability of **not** rolling a 6 is $1 - \frac{1}{6} = \frac{5}{6}$

P(Red) + P(Yellow) + P(Green) = 1

---

## Worked example

A spinner can land on red, blue, white or green. The table shows the probabilities of the spinner landing on each colour.

| Colour | Red | Blue | White | Green |
|---|---|---|---|---|
| Probability | 0.12 | $x$ | $2x$ | 0.28 |

Work out the value of $x$. **(3 marks)**

$$0.12 + x + 2x + 0.28 = 1$$
$$0.4 + 3x = 1 \quad (-0.4)$$
$$3x = 0.6 \quad (\div 3)$$
$$x = 0.2$$

## Problem solved!

The probabilities have to add up to 1. Use this fact to write an equation, then solve it to work out the value of $x$.

For a reminder about solving linear equations have a look at page 30.

Check:
P(Blue) = $x$ = 0.2     P(White) = $2x$ = 0.4
0.12 + 0.2 + 0.4 + 0.28 = 1 ✓

You will need to use problem-solving skills throughout your exam – **be prepared!**

---

## Expectation

If you flip a coin 100 times, you can expect to get heads about 50 times.
You probably won't get heads exactly 50 times, but it's a good guess.

Expected number of outcomes = number of trials × probability

You can use expectation to help you decide if a dice or coin is **fair**.
These two coins have been flipped 50 times each.

Coin 1

| Head | ЖЖ ЖЖ ЖЖ ЖЖ ||| |
|---|---|
| Tail | ЖЖ ЖЖ ЖЖ ЖЖ ЖЖ || |

About the same number of heads and tails.
This coin is probably **fair**.

Coin 2

| Head | ЖЖ ЖЖ |||| |
|---|---|
| Tail | ЖЖ ЖЖ ЖЖ ЖЖ ЖЖ ЖЖ ЖЖ | |

A lot more than the expected number of tails.
This coin is probably **biased**.

---

Worked solution video

## Now try this

Preti has a packet of sweets. The sweets are red, yellow, green or orange.

Preti picks one sweet at random. This table gives the probability that the sweet will be red, yellow, or orange.

| Colour | Red | Yellow | Green | Orange |
|---|---|---|---|---|
| Probability | 0.13 | 0.36 | | 0.28 |

Complete the table by working out the probability that the sweet will be green.

**(2 marks)**

# Relative frequency

You need to be able to calculate probabilities for data given in graphs and tables. You can use this formula to estimate a probability from a frequency table:

$$\text{Probability} = \frac{\text{frequency of outcome}}{\text{total frequency}}$$

When a probability is calculated like this it is sometimes called a **relative frequency**.

### Golden rule

Probability estimates based on relative frequency are **more accurate** for larger samples (or for more trials in an experiment).

In the sample there were 15 + 10 = 25 eggs which weighed 55 g or more. So an estimate for the probability of picking an egg that weighs 55 g or more is $\frac{25}{40}$ or $\frac{5}{8}$.

An egg farm weighed a sample of 40 eggs. It recorded the results in a frequency table.

| Weight, $w$ (g) | Frequency |
|---|---|
| $45 \leqslant w < 50$ | 6 |
| $50 \leqslant w < 55$ | 9 |
| $55 \leqslant w < 60$ | 15 |
| $60 \leqslant w < 65$ | 10 |

(a) Roselle buys some eggs from the farm and picks one at random. Estimate the probability that the egg weighs 55 g or more. **(2 marks)**

$$P(w \geqslant 55) \approx \frac{25}{40} = \frac{5}{8}$$

(b) Comment on the accuracy of your estimate. **(1 mark)**

40 is a fairly small sample size, so the estimate is not very accurate.

## Experimental probability

You can carry out an experiment to estimate the probability of something happening. This table shows the results of throwing a drawing pin 60 times.

| Number of trials | 10 | 20 | 30 | 40 | 50 | 60 |
|---|---|---|---|---|---|---|
| Frequency of landing point up | 8 | 11 | 17 | 25 | 30 | 37 |

To estimate the probability that the drawing pin will land point up, you calculate the relative frequency. The most accurate estimate will be based on the largest number of trials.

A four-sided dice is rolled 40 times. The results are shown in the table.

| Number | 1 | 2 | 3 | 4 |
|---|---|---|---|---|
| Frequency | 8 | 4 | 21 | 7 |

(a) Work out the estimated probability of getting a 3. **(1 mark)**

(b) Work out the theoretical probability of getting a 3 on a fair four-sided dice. **(1 mark)**

(c) Do you think this dice is fair? Give a reason for your answer. **(1 mark)**

Worked solution video

If a dice is fair, the experimental probability gets closer to the theoretical probability as the number of trials increases.

# Frequency and outcomes

You might need to record all the possible **outcomes** of two or more events. You can use a **frequency tree** to show the frequencies of each possible outcome.

## Worked example

Wilfred owns 80 books. He has **not** read 15 of these books.

20 of the books he **has** read are hardbacks. He has 52 paperbacks in total. The rest of the books are hardbacks.

(a) Use this information to complete this frequency tree.

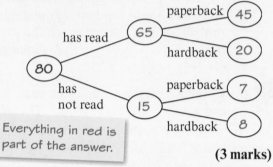

*Everything in red is part of the answer.*

**(3 marks)**

Wilfred chooses one of his books at random.

(b) Work out the probability that it is a paperback he has not read.    **(2 marks)**

$\frac{7}{80}$

## Golden rule

In a frequency tree, each frequency is equal to the sum of its branches.

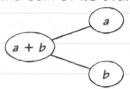

1.  Write 15 in the bottom left oval, then use the golden rule above. Wilfred has read $80 - 15 = 65$ books.
2.  He has $80 - 52 = 28$ hardbacks in total. 20 of these are books he **has** read, so 8 must be books he has **not** read.
3.  Use the golden rule to complete the frequency tree.

## Sample space diagrams

A **sample space diagram** shows you all the possible outcomes of two events.

Here are all the possible outcomes when two coins are flipped.

There are four possible outcomes. TH means getting a tail on the first coin and a head on the second coin.

## Worked example

This bag contains 30 counters.
They are all either black or white.
A counter is chosen at random.
The probability that it is black is $\frac{1}{5}$. How many white counters are in the bag?    **(2 marks)**

$P(\text{White}) = 1 - \frac{1}{5} = \frac{4}{5}$

$\frac{4}{5} \times 30 = 24$

**Check it!**
There are $30 - 24 = 6$ black counters in the bag.
$P(\text{Black}) = \frac{6}{30}$ ✔

## Now try this

Dawn is eating a meal at a restaurant. She is in a rush so she chooses one main course and one dessert from the menu at random.

(a) Write down all the possible outcomes.  **(2 marks)**

(b) Work out the probability that Dawn chooses fish pie and cheesecake.    **(3 marks)**

**Main courses**
Steak and chips
Fish pie
Mushroom pasta

**Desserts**
Cheesecake
Lemon mousse
Tiramisu

# Venn diagrams

You can use a Venn diagram to show frequencies in a probability question. This Venn diagram shows the results when 50 people were asked whether they owned a dog (*D*) or a cat (*C*). The rectangle represents everyone who was surveyed. The **number** in each section tells you **how many** people that section represents.

In total 21 + 8 + 6 + 15 = 50 people were surveyed. This symbol represents all of them.

Some people owned a cat *and* a dog so the ovals overlap.

This oval represents people who owned a cat.

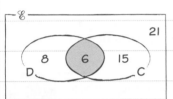

6 people owned a dog **and** a cat. You can write this as $D \cap C$. ∩ means **and** or **intersection**.

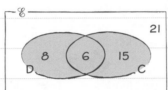

8 + 6 + 15 = 29 people owned a dog **or** a cat. You can write this as $D \cup C$. ∪ means **or** or **union**.

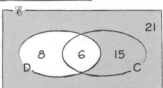

15 + 21 = 36 people **did not** own a dog. You can write this as $D'$. $D'$ means **not** *D* or the **complement** of *D*.

---

## Worked example

36 members of a youth club were surveyed about the sports they played.

19 members played tennis.

14 members played football.

6 members played both tennis and football.

(a) Draw a Venn diagram to show this information. **(3 marks)**

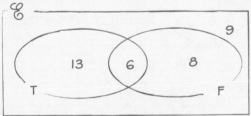

(b) One of the members is chosen at random. Write down the probability that the member plays neither tennis nor football. **(1 mark)**

$\frac{9}{36} = \frac{1}{4}$

$T \cap F$

Fill in the centre of the Venn diagram first. 6 members play both sports.

$T \cap F'$

The 19 members who play tennis include the 6 members who play both sports. So 19 − 6 = 13 members play tennis but *not* football.

$T' \cap F$

14 − 6 = 8 members play football but *not* tennis.

$T' \cap F'$

36 − 13 − 6 − 8 = 9 members play neither sport.

**Check it!**

13 + 6 = 19 members play tennis ✓

8 + 6 = 14 members play football ✓

13 + 6 + 8 + 9 = 36 members surveyed ✓

---

## Now try this

The Venn diagram shows information about how 30 people in an office travel to work.

One person is chosen at random. Find the probability that this person

(a) uses both the bus and the train on their journey **(1 mark)**

(b) uses the bus at some point in their journey. **(2 marks)**

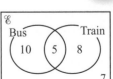

# Independent events

Two events are **independent** if the outcome of one does not affect the outcome of the other. You can work out the probability of two independent events **both** occurring by **multiplying** the probabilities.

## Worked example

Huan spins this spinner. He keeps spinning until he lands on green.

Work out the probability that Huan spins the spinner

(a) exactly twice                          (2 marks)

P(blue) × P(green) = $\frac{7}{8}$ × $\frac{1}{8}$ = $\frac{7}{64}$

(b) more than twice.                    (2 marks)

P(blue) × P(blue) = $\frac{7}{8}$ × $\frac{7}{8}$ = $\frac{49}{64}$

## Problem solved!

(a) Huan stops when he lands on green, so in order to spin the spinner **exactly** twice he must land on blue on the first spin, **and** green on the second spin.

(b) In order for Huan to spin more than twice he must land on blue on **both** of the first two spins.

You will need to use problem-solving skills throughout your exam – **be prepared!**

## Tree diagrams

You can show independent events on a tree diagram.

You write the probabilities on the branches.
For each pair of branches, the probabilities add up to 1:
$\frac{1}{8} + \frac{7}{8} = 1$

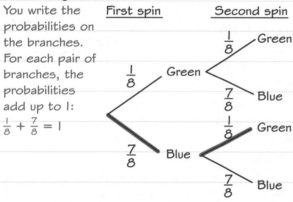

Work out probabilities using these rules:

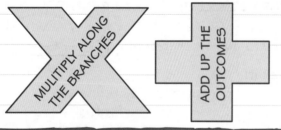

MULTIPLY ALONG THE BRANCHES

ADD UP THE OUTCOMES

## Worked example

Aidan and Chloe each take a basketball shot.
The probability that Aidan scores is 0.3
The probability that Chloe scores is 0.4
Work out the probability that exactly one player scores.                    (3 marks)

P(Exactly one player scores)
   = 0.3 × 0.6 + 0.7 × 0.4
   = 0.18 + 0.28
   = 0.46

There are **two** successful outcomes. Multiply along the branches to work out the probability of each one, then add up the answers.

## Now try this

Ravina and Anjali are going to school one morning.
The probability that Ravina will arrive late is 0.2.
The probability that Anjali will arrive late is 0.5.

(a) Draw a probability tree diagram to show this information.                    (3 marks)

(b) Work out the probability that Ravina and Anjali will both arrive late for school.                    (2 marks)

# Problem-solving practice 1

About half of the questions in your Foundation GCSE exam will require you to **problem-solve, reason, interpret** or **communicate** mathematically. If you come across a tricky or unfamiliar question in your exam you can try some of these strategies:

- ✓ Sketch a diagram to see what is going on.
- ✓ Try the problem with smaller or easier numbers.
- ✓ Plan your strategy before you start.
- ✓ Write down any formulae you might be able to use.
- ✓ Use *x* or *n* to represent an unknown value.

**1** The table shows information about the numbers of Year 7 pupils absent from Keith's school last week.

|           | Boys | Girls |
|-----------|------|-------|
| Monday    | 8    | 10    |
| Tuesday   | 11   | 9     |
| Wednesday | 12   | 12    |
| Thursday  | 14   | 13    |
| Friday    | 13   | 11    |

Keith wants to compare the data.
Draw a suitable diagram or chart.   **(4 marks)**

**Bar charts page 118**
**Line graphs page 124**

In this question **you** have to choose what type of diagram or chart to use. It is best to use a bar chart or a line graph.

**TOP TIP**
- Label **both** axes correctly.
- Draw a key for boys and girls, or make sure it is clear which bars (or lines) are for boys and which are for girls.

**2** Abi has five cards.
Each card has a number written on it.

The mean of the five numbers is 6.
One of the numbers is hidden.
Work out the hidden number. **(2 marks)**

**Averages and range page 121**

You could try some different values for the hidden number and work out the mean each time. But you can save time by using the rule in the Top tip. There are five numbers and the mean is 6, so the sum of the numbers must be 5 × 6 = 30.

**TOP TIP**
mean × number of data values = sum of data values

# Problem-solving practice 2

 **3** Some students in a class weighed themselves.

Here are their results.

Boys' weights in kg
70  65  45  52  63  72  63

Girls' weights in kg
65  45  47  61  44  67  55  56  63

Compare fully the weights of these students.               (6 marks)

### Averages and range page 121

There are **6 marks** for this question.
To give a full answer you need to **compare** the data. So (1) calculate averages like the mean or median and a measure of spread like the range, and (2) write a sentence for each of these, comparing the boys and the girls.

**TOP TIP**

You need to **calculate** first, then use your calculations to write your **conclusion**. It's safest to write in full sentences for your conclusion.

---

**4** The scatter graph shows the French and German marks of 15 students.

French mark

Jade's French mark was 42.

Estimate Jade's German mark.     (2 marks)

### Scatter graphs page 120

Start by drawing a line of best fit on the scatter graph. Next draw a line up from 42 on the horizontal axis to your line of best fit. Then draw a line across to the vertical axis to estimate Jade's German mark. Make sure you draw the lines you use on your graph to show your working.

**TOP TIP**

When you are reading information from a graph you should always give your answer to the nearest small square.

---

 **5** Amy picks one card at random from group A and one card at random from group B.

A                           B

She adds together the numbers on the two cards. What is the probability that her total is
(a) 8      (b) 7      (c) an odd number?
                                (4 marks)

### Independent events page 132
### Counting strategies page 19

You can tackle this in two different ways.
1. Write down all the possible outcomes when 2 cards are picked.
2. Consider the two picks as two separate independent events, and multiply the probabilities of each.

**TOP TIP**

In a probability question make sure you have considered all the possible outcomes, and all the successful outcomes.

# Answers

## NUMBER

### 1. Place value

1. £974, £1497, £1749, £1947, £1974
2. (a)

| Car | A | B | C | D | E |
|---|---|---|---|---|---|
| Bought | £2200 | £7800 | £4200 | £10700 | £6500 |
| Sold | £2900 | £9000 | £2700 | £11200 | £5750 |
| Profit or loss | £700 profit | £1200 profit | £1500 loss | £500 profit | £750 loss |

(b) £750 profit

### 2. Negative numbers

1. (a) –6     (b) 5
2. (a) 48     (b) –6
3. –3 and –5, –10 and 2, –9 and 1

### 3. Rounding numbers

1. (a) 3800     (b) 3760
2. (a) 50     (b) 3000     (c) 0.009     (d) 7
3. 0.180
4. No. The cooker could be more than 76 cm wide (e.g. 76.3 cm).

### 4. Adding and subtracting

1. (a) 1714     (b) 1028
2. (a) 376     (b) 172
3. 221 and 605, 492 and 334, 552 and 274
4. £1.57

### 5. Multiplying and dividing

1. (a) 7200     (b) 2304     (c) 2262
2. (a) 156     (b) 129
3. $18 \times 6 = 108$, $7 \times 15 = 105$. Yes, she has enough coins.

### 6. Decimals and place value

1. (a) 8 tenths     (b) 8 thousandths
2. 0.508, 0.51, 0.517, 0.571, 0.58
3. (a) 8.736     (b) 130

### 7. Operations on decimals

1. (a) 20.43     (b) 11.912
2. (a) 6.78     (b) 195.84
3. (a) £239.40     (b) 94p

### 8. Squares, cubes and roots

1. (a) 36     (b) 5     (c) 64     (d) 11
2. No. e.g. $25 = 5^2$, $9 = 3^2$, $25 \times 9 = 225$ which is not even

### 9. Indices

1. (a) $5^{10}$     (b) $5^7$     (c) $5^3$     (d) $5^{15}$
2. (a) $7^5$     (b) 64
3. (a) 1    (b) $\frac{1}{4}$    (c) $\frac{1}{16}$    (d) $\frac{4}{25}$    (e) $\frac{64}{27} = 2\frac{10}{27}$

### 10. Estimation

$$\frac{80 \times 300}{60 \times 40} = 10$$

(a) 108 (or 81, rounding $\frac{4}{3}$ to 1)

(b) $3 < 3.14$ and $3 < 3.2$ (and $1 < \frac{4}{3}$) so underestimate

### 11. Factors, multiples and primes

1. (a) 9     (b) 20     (c) 24
2. $280 = 2^3 \times 5 \times 7$

### 12. HCF and LCM

1. (a) $132 = 2^2 \times 3 \times 11$, $110 = 2 \times 5 \times 11$
   (b) 22     (c) 660
2. (a) $3^2 \times 5$     (b) $2 \times 3^3 \times 5^2 \times 7^4$

### 13. Fractions

1. £8 profit
2. e.g. $10 \times 0.25 + 20 \times 0.3 = 8.5$, so Sandeep makes a 50p profit

### 14. Operations on fractions

1. (a) $\frac{8}{9}$     (b) $\frac{9}{16}$
2. (a) $\frac{3}{8}$     (b) $\frac{7}{55}$
3. (a) $\frac{1}{6}$     (b) $\frac{14}{15}$

### 15. Mixed numbers

1. (a) $5\frac{5}{6}$     (b) $2\frac{6}{25}$
2. 7

### 16. Calculator and number skills

1. (a) 4     (b) 7     (c) 17     (d) 125
2. (a) 31     (b) 1.4

### 17. Standard form 1

1. (a) $2.8 \times 10^{-4}$     (b) 391 000
2. (a) $2.71 \times 10^6$     (b) $1.75 \times 10^7$

### 18. Standard form 2

1. (a) $2.7 \times 10^{12}$     (b) $3.2 \times 10^{-6}$

### 19. Counting strategies

1. AY, AZ, BY, BZ, CY, CZ
2. 15

### 20. Problem-solving practice 1

1. 10 p
2. e.g.
   Cost of theatre trip
   £15 × 25 + £11.50 × 5 + £5.75 × 20 = £547.50
   Cost of zoo trip
   £240 + £18 × 5 + £12 × 20 = £570
   Conclusion
   The lowest possible total cost is the theatre with circle tickets: £547.50

### 21. Problem-solving practice 2

3. $\frac{7}{20}$
4. $2 \times \frac{3}{8} = \frac{3}{4}$ of a bag per day.
   $14 \div \frac{3}{4} = 18\frac{2}{3}$. Susan can feed the dogs for 18 days from 1 bag.
5. 96 mm

# ALGEBRA

## 22. Collecting like terms

1. (a) $6x$    (b) $8q$    (c) $10t$    (d) $3n$
2. term
3. $9p - 3q$

## 23. Simplifying expressions

(a) $n^5$    (b) $6r^3$    (c) $32yz$    (d) $4f$    (e) $7$

## 24. Algebraic indices

1. (a) $m^9$    (b) $k^7$
2. (a) $h^8$    (b) $t^{20}$    (c) $a^5$
3. (a) $y^8$    (b) $125p^6$

## 25. Substitution

1. 12
2. 6
3. (a) 13    (b) 51    (c) 23

## 26. Formulae

1. $36\,\text{cm}^2$
2. 49

## 27. Writing formulae

1. $T = 0.5x + 0.35y$
2. (a) $A = 3s^2$    (b) $75\,\text{cm}^2$

## 28. Expanding brackets

1. (a) $3y - 18$    (b) $m^2 + 7m$
2. (a) $13a - 6b$    (b) $-w + 21$    (c) $m^2 + 22m$

## 29. Factorising

1. (a) $6(n + 3)$    (b) $3(p - 1)$
2. (a) $4t(t + 2)$    (b) $3x(x - 4)$
3. $2, 2p, p - 2q$

## 30. Linear equations 1

1. (a) $y = 8$    (b) $m = 17$    (c) $k = 19$
2. (a) $n = 5$    (b) $w = 10.5$
3. (a) $n = 36$    (b) $q = 77$

## 31. Linear equations 2

1. (a) $x = 5$    (b) $x = 0.5$
2. (a) $t = -11$    (b) $h = -0.5$
3. (a) $y = 14.5$    (b) $m = -1$

## 32. Number machines

12

## 33. Inequalities

1. (a) $x > -3$    (b) $x \leqslant 4$
2. (a)

(b)

## 34. Solving inequalities

1. (a) $x \geqslant 5$    (b) $4 > x$ (or $x < 4$)
2. (a) $-2, -1, 0, 1, 2$    (b) $0, 1, 2, 3$

## 35. Sequences 1

1. 33, 45
2. (a) 51
   (b) 11th term (= 72)
3. 21, 34, 55

## 36. Sequences 2

1. (a) $n$th term $= 4n - 1$
   (b) No. 22nd term $= 87$, 23rd term $= 91$
      (or $4n - 1 = 89$
              $n = 22.5$ which is not a whole number)
2. (a) $-3n + 29$    (b) $-1$

## 37. Coordinates

(a) $(3, 6)$
(b) 6 units
(c) $(6, 4)$

## 38. Gradients of lines

(a) 5
(b) No. The gradient is 5 so for every £10 000 spent revenue increases by approximately £50 000.

## 39. Straight-line graphs 1

$y = -2x + 2$

## 40. Straight-line graphs 2

1. $y = 6x - 20$
2. $y = 2x - 1$

## 41. Real-life graphs

(a) £5
(b) £15
(c) Gradient $= 2.5$ so additional cost per minute $= 2.5$p. So additional 10 minutes cost 25p. Claim is true.

## 42. Distance–time graphs

(a) 30 minutes
(b) 30 km/h

## 43. Rates of change

A: False. The car was accelerating between 0 and 5 seconds.
B: True: The graph is flat between 5 and 15 seconds.
C: True: The graph slopes upwards 0 and 5 seconds.
D: False: The car was travelling at a constant speed between 5 and 15 seconds.

## 44. Expanding double brackets

1. $(x + 6)(x + 3) = x^2 + 3x + 6x + 18$
                $= x^2 + 9x + 18$
2. (a) $x^2 + 4x - 5$
   (b) $p^2 - 12x + 36$

## 45. Quadratic graphs

(a)

| $x$ | $-3$ | $-2$ | $-1$ | $0$ | $1$ | $2$ | $3$ |
| --- | --- | --- | --- | --- | --- | --- | --- |
| $y$ | 11 | 6 | 3 | 2 | 3 | 6 | 11 |

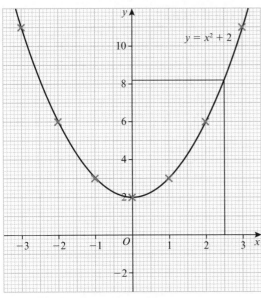

$y = 8.2$ (to the nearest small square)

## Using quadratic graphs
$x = -0.2$ and $x = 4.2$
$x = -1.6$ and $x = 5.6$

## Factorising quadratics
(a) $(x + 2)(x + 4)$      (b) $(x - 2)(x - 8)$
(a) $(x - 12)(x + 12)$      (b) $(x + 7)(x - 7)$

## Quadratic equations
(a) $(x + 3)(x + 2)$      (b) $x = -2, x = -3$
(a) $x = 0, x = 5$
(b) $x = 4, x = -7$
(c) $x = 12, x = -12$

## Cubic and reciprocal graphs
(a) D    (b) E    (c) A    (d) B    (e) F    (f) C

## Simultaneous equations
$x = 5, y = 1.5$

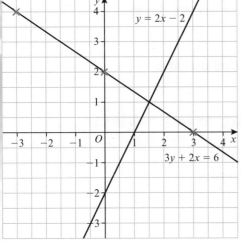

$x = 1.5, y = 1$

## Rearranging formulae
(a) $Q = \dfrac{W + 100}{7}$      (b) $v = \dfrac{m}{d}$

$n = \dfrac{2m - p}{2}$

## 52. Using algebra
1. $(x - 3) + 2x + (x - 3) + 2x = x + (x + 4) + x + (x + 4)$
$$6x - 6 = 4x + 8$$
$$2x = 14$$
$$x = 7$$
Rectangle A: width $= 4\,$m, height $= 14\,$m
2. No. e.g. $n = 4$: $n^2 + 4n + 3 = 35 = 5 \times 7$

## 53. Identities
1. $(a + b)(a - b) = a^2 - ab + ab - b^2$
$$= a^2 - b^2$$
2. $(n + 2)(n - 2)$
3. True, False, False, True

## 54. Problem-solving practice 1
1. e.g. $A(10, 10)$, $B(50, 10)$, $C(50, 30)$, $D(10, 30)$
2. $1.4\,$kg

## 55. Problem-solving practice 2
3. $t = 3.5$
4. $n$th term $= 6n + 3$
$$6 \times 10 + 3 = 63$$
$$6 \times 11 + 3 = 69$$
So 65 is not a term in the sequence.
5. 3 minutes $= \frac{3}{60} = \frac{1}{20}$ hours
Speed $= \frac{\text{distance}}{\text{time}} = y \div \frac{1}{20} = y \times \frac{20}{1} = 20y$
6. (a) $y = 8x - 30$
e.g. When $x = 10$: $y = 8 \times 10 - 30 = 50$, so $(10, 50)$ is a point on the line.
(b) e.g. The answer is reliable because it uses algebra

# RATIO & PROPORTION

## 56. Percentages
1. £546
2. 35%

## 57. Fractions, decimals and percentages
1. (a) $\frac{3}{20}$      (b) $\frac{17}{25}$
2. 36%, $\frac{2}{5}$, 0.42
3. $5 \times 12 = 60$ counters in total
35% of $60 = 21$
$\frac{2}{5}$ of $60 = 24$
$60 - 21 - 24 = 15$
Haydon won 15 counters

## 58. Percentage change 1
1. 1351
2. 34.6% (1 d.p.)

## 59. Percentage change 2
Cruks cameras price $=$ £171.50
Spivs cameras price $=$ £175.50
Cruks cameras is cheaper.

## 60. Ratio 1

1. (a) $\frac{4}{7}$ (b) $2:3$
2. $7:10$
3. 56 counters

## 61. Ratio 2

1. £120
2. Rice : Onion = $300:160 = 15:8$

## 62. Metric units

1. (a) 320 cm (b) 250 ml (c) 96 cm (d) 1.7 kg
2. 1.7 mm

## 63. Reverse percentages

1. £45
2. £220 000

## 64. Growth and decay

(a) £5304.50 (b) 4 years

## 65. Speed

1. 4 hours and 15 minutes
2. 266 miles

## 66. Density

1. $300 \div 7.8$
2. 9.234 g/cm$^3$

## 67. Other compound measures

1. 16.6 km/l
2. 2.4 minutes or 2 minutes and 24 seconds

## 68. Proportion

1. (a) £2.80 (b) £6.72
2. 16 days

## 69. Proportion and graphs

(a) 32 cm (b) 40 inches
(c) Graph is a straight line and passes through the origin (0, 0)

## 70. Problem-solving practice 1

1. 45
2. Sportcentre: £45  Footwear First: £60  Action Sport: £48
   Sportcentre is the cheapest.

## 71. Problem-solving practice 2

3. 312
4. £3610.80
5. 24.72 litres

# GEOMETRY & MEASURES

## 72.  Symmetry

(a) e.g. 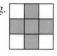 (b) e.g.

## 73. Quadrilaterals

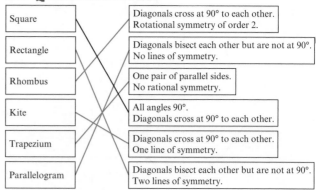

## 74. Angles 1

$b = 55°$ (Vertically opposite angles are equal)
$c = 125°$ (Angles on a straight line add up to 180°)
$d = 125°$ (Vertically opposite angles are equal)

## 75. Angles 2

(a) $x = 134°$ (Corresponding angles are equal)
(b) $y = 97°$

## 76. Solving angle problems

1. (a) 71° (b) 24°
2. $56 + y + (y + 60) = 180$
   $2y + 116 = 180$
   $2y = 64$
   $y = 32$
   Angle $y = 32°$

## 77. Angles in polygons

144°

## 78. Time and timetables

(a) 14:36 (b) 8 hours

## 79. Reading scales

25 g

## 80. Perimeter and area

34 cm$^2$

## 81. Area formulae

$x = 8$

## 82. Solving area problems

(a) 3 cm (b) 5 cm (c) 58 cm$^2$ (d) 50 cm

## 83. 3D shapes

(a) 250 cm$^2$
(b) $50 \times 250 = 12\,500$
   $12\,500 \div 3000 = 4\frac{1}{6}$
   Karl must buy 5 cans.

## 84. Volumes of cuboids

6.5 cm

## 85. Prisms

(a) 1200 cm$^3$ (b) 920 cm$^2$

# 86. Units of area and volume

(a) 23 000 cm² (b) 0.4 cm³

3.5 × 10⁸ mm³

37 500 N/m²

# 87. Translations

(a)

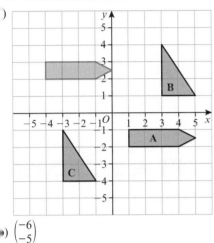

(b) $\begin{pmatrix} -6 \\ -5 \end{pmatrix}$

# 88. Reflections

(a), (b), (c)

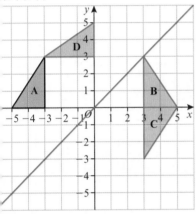

(d) Reflection in the line $y = -x$

# 89. Rotations

(a) Rotation 90° anticlockwise about centre (4, 3)

(b) Rotation 180° about centre (6, 3)

# 90. Enlargements

(a), (b)

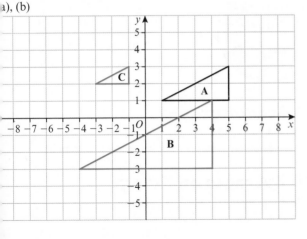

# 91. Pythagoras' theorem

(a) $y = 3.5$ cm (b) $z = 9.1$ cm

# 92. Line segments

1. 9.22 (2 d.p.)
2. 18.56 (2 d.p.)

# 93. Trigonometry 1

(a) 44.8° (b) 56.2° (c) 48.6°

# 94. Trigonometry 2

(a) 4.4 cm (b) 5.4 cm (c) 15.7 cm

# 95. Solving trigonometry problems

$\cos 60° = \dfrac{2.4}{AC}$

$AC \times \cos 60° = 2.4$

$AC = \dfrac{2.4}{\cos 60°}$

$= 2.4 \div \frac{1}{2} = 4.8$ m

# 96. Measuring and drawing angles

1. (a) 28° (b) 308°
2.

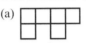

# 97. Measuring lines

1. (a) 2.8 cm (b) 5.5 cm (c) 4.9 cm
   (d) 6.0 cm
2. Accept between 20 and 25 feet

# 98. Plans and elevations

(a)  (b) (c)

(d) 30 cm²

# 99. Scale drawings and maps

80 cm

# 100. Constructions 1

# 101. Constructions 2

1.

2.

## 102. Loci

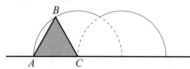

## 103. Bearings

(a) 280 km (accept 275 km to 290 km)
(b) 225°
(c)

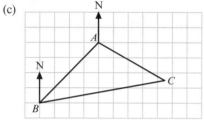

## 104. Circles

(a) 38.2 cm
(b) 19.1 cm

## 105. Area of a circle

9.4 cm

## 106. Sectors of circles

12.6 cm (3 s.f.)

## 107. Cylinders

55 litres

## 108. Volumes of 3D shapes

Volume of cone $= \frac{1}{3}\pi r^2 h = 3\pi h$
Volume of sphere $= \frac{4}{3}\pi r^3 = 36\pi$
$3\pi h = 36\pi$ $(\div 3\pi)$
$h = 12$ cm

## 109. Surface area

1. $52\pi$
2. $115\,\text{cm}^2$ (3 s.f.)

## 110. Similarity and congruence

A and K, H and J, D and G

## 111. Similar shapes

(a) 130° (b) 31.5 cm

## 112. Congruent triangles

Angle $ABC$ = Angle $ADC$ = 90°
Hypotenuse of both triangles are a common side ($AC$) so are equal
$BC = CD$
So RHS is satisfied and triangles are congruent.

## 113. Vectors

(a) $\begin{pmatrix} 6 \\ 10 \end{pmatrix}$ (b) $\begin{pmatrix} 5 \\ -4 \end{pmatrix}$ (c) $\begin{pmatrix} -7 \\ -24 \end{pmatrix}$

## 114. Problem-solving practice 1

1. $10y + 90° = 360°$
   $y = 27°$
2. 13.8 km (1 d.p.)

## 115. Problem-solving practice 2

3. e.g. Radius of large semicircle = $2x$
   Radius of small semicircle = $x$
   Shaded area $= \frac{1}{2}\pi x^2 + \frac{1}{2}\pi x^2 = \pi x^2$
   Area of large semi-circle $= \frac{1}{2}\pi(2x)^2 = 2\pi x^2$
   Which is twice the shaded area
4. 5.8 m (1 d.p.)
5. Interior angle of A = 360° − 90° − 108° = 162°
   Exterior angle of A = 18°
   $n = 360° \div 18° = 20$

# PROBABILITY & STATISTIC

## 116. Two-way tables

|  | Gloss | Matt | Lustre | Total |
|---|---|---|---|---|
| **Small** | 20 | 35 | 12 | 67 |
| **Medium** | 63 | 102 | 29 | 194 |
| **Large** | 22 | 24 | 18 | 64 |
| **Total** | 105 | 161 | 59 | 325 |

## 117. Pictograms

(a)

Key: = £50 000

| Lancashire | |
|---|---|
| Herefordshire | |
| Essex | |
| Surrey | |
| Lincolnshire | |

(b) £250 000

## 118. Bar charts

56 applications

## 119. Pie charts

Salad: 84°, Stew 36°, Pizza: 180°, Curry: 60°

## 120. Scatter graphs

(a) Accept 23.5° to 24.5°
(b) This is not a reliable estimate because it is extrapolation

## 121. Averages and range

Mode: 14, Median: 13, Mean: 12, Range: 7
'The mean is larger than the mode' is incorrect

## 122. Averages from tables 1

(a) 1 goal (b) 2 goals (c) $95 \div 50 = 1.9$ goals

## 123. Averages from tables 2

(a) $10 < x \leqslant 20$    (b) $20 < x \leqslant 30$    (c) $1350 \div 50 = 27$

## 124. Line graphs

(a)

(b) Upward trend (or upward trend from 2009 onwards)

## 125. Sampling

(a) Amy only observed 6 trains so her statement is unreliable.

(b) $\frac{8}{25} = 0.32$

## 126. Comparing data

Anna's mean $104 \div 8 = 13$   Anna's range $16 - 9 = 7$

Carla's mean $11$    Carla's range $14 - 10 = 4$

On average, Anna scored better than Carla because she had a higher mean.

Carla's scores were more consistent than Anna's because she had a smaller range.

## 127. Probability 1

(a) $\frac{2}{10}$ or $\frac{1}{5}$    (b) $\frac{5}{10}$ or $\frac{1}{2}$    (c) $0$

## 128. Probability 2

| Colour | Red | Yellow | Green | Orange |
|---|---|---|---|---|
| Probability | 0.13 | 0.36 | 0.23 | 0.28 |

## 129. Relative frequency

(a) $\frac{21}{40}$      (b) $\frac{1}{4}$

(c) No. $\frac{21}{40}$ is more than twice $\frac{1}{4} = \frac{10}{40}$ so the dice is not fair.

## 130. Frequency and outcomes

(a) SC, SL, ST, FC, FL, FT, MC, ML, MT

(b) $\frac{1}{9}$

## 131. Venn diagrams

(a) $\frac{5}{30} = \frac{1}{6}$    (b) $\frac{15}{30} = \frac{1}{2}$

## 132. Independent events

(a)

(b) P(Both late) $= 0.2 \times 0.5 = 0.1$

## 133. Problem-solving practice 1

1.

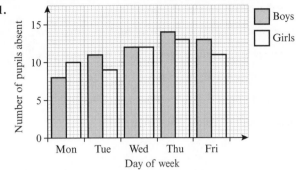

2. 9

## 134. Problem-solving practice 2

3. e.g.

| | Mean | Median | Range |
|---|---|---|---|
| **Boys** | 61.4 kg | 63 kg | 27 kg |
| **Girls** | 55.9 kg | 56 kg | 23 kg |

The boys had a higher mean and median than the girls, so they were on average heavier. The girls had a smaller range which indicates that their weights were more consistent than the boys' weights.

4.

Estimated German mark is 47

5. (a) $\frac{1}{6}$      (b) $\frac{2}{6} = \frac{1}{3}$      (c) $\frac{3}{6} = \frac{1}{2}$

Published by Pearson Education Limited, 80 Strand, London, WC2R 0RL.

www.pearsonschoolsandfecolleges.co.uk

Copies of official specifications for all AQA qualifications may be found on the AQA website: www.aqa.org.uk

Text © Harry Smith and Pearson Education Limited 2016

Typeset and illustrations by Tech-Set Ltd, Gateshead

Original illustrations © Pearson Education Limited 2016

Produced by Out of House Publishing

Cover illustration by Miriam Sturdee

The right of Harry Smith to be identified as author of this work has been asserted by him in accordance with the Copyright, Designs and Patents Act 1988.

First published 2016

22

10 9 8

**British Library Cataloguing in Publication Data**
A catalogue record for this book is available from the British Library

ISBN 9781447988069

Printed in Great Britain by Bell and Bain Ltd, Glasgow

**Note from the publisher**
Pearson has robust editorial processes, including answer and fact checks, to ensure the accuracy of the content in this publication, and every effort is made to ensure this publication is free of errors. We are, however, only human, and occasionally errors do occur. Pearson is not liable for any misunderstandings that arise as a result of errors in this publication, but it is our priority to ensure that the content is accurate. If you spot an error, please do contact us at resourcescorrections@pearson.com so we can make sure it is corrected.